Experimental Bases for Chemical Theory

Experimental Bases for Chemical Theory

Experimental Bases for Chemical Theory

P. D. Arculus
Solihull School

M. J. English
Cannock Grammar School

Oliver and Boyd
Edinburgh and London

OLIVER & BOYD LTD
Tweeddale Court, Edinburgh 1
39a Welbeck Street, London W.1

First Published 1968
05 001517 6

Printed in Great Britain by
T. & A. Constable Ltd, Edinburgh

PREFACE

The contents of this book are suitable for candidates at the Advanced Level of this General Certificate of Education and the Ordinary National Certificate, as well as for students at colleges of education and their first year at university.

The experiments do not demand elaborate apparatus and are not expensive to carry out. Each one is designed to be performed either by an individual working on his own or, better, simultaneously by all members of a practical group, with each student doing the same experiment as his colleagues. Although we appreciate that the latter alternative depends upon the facilities available, we have never been forced to restrict the performance of any experiment to a few students.

We have conducted all the experiments in a variety of situations and under different conditions, have been greatly helped by the valuable advice and encouragement given to us by Mr. P. H. Arnold, B.Sc., and are appreciative of the provisions in the Chemistry Department of Solihull School and of the assistance of many members of the Science Sixth Forms of the School, in particular Mr. C. J. Chapman, Mr. P. J. F. Millier, Mr. J. C. W. Richards, Mr. H. J. C. Taylor and Mr. R. J. Wyatt.

Experiments may be selected from each of the various chapters of this book to provide a sound practical basis for topics other than those usually associated with traditional qualitative analysis and organic chemistry. Such a selection would not be a complete practical course: teachers will realise that easily performed test tube reactions are an essential preliminary, at least, to all the experiments.

Although we have not found it easy to convey them as we would wish, our views on the teaching of chemistry are implicit in the presentation of the experiments. The minimum of theory has been included to enable the student to understand the principles involved in any experiment. After what are, we trust, clear and accurate statements of requirements and instructions, suggestions are made on the recording and interpretation of observations and on the formulation of conclusions. We hope the questions, at the end of the details for each experiment, will encourage students to think for themselves, to enter into discussion, and to seek further information from both data tables and the standard textbooks of inorganic and physical chemistry. The questions are an important feature of the book: they arise directly from the practical work and often provide an extension to it. The answers given to the questions are equally important, for they should reveal either understanding or a lack of it! We hope that this collection of experiments and the questions associated with them will lead to logical practice and sensible thinking in a subject which relies on the experimental approach.

We are indebted to many members of the Association for Science Education and the Royal Institute of Chemistry who, either through their contributions to the *School Science Review* and *Education in Chemistry*, or by correspondence, or in discussion, have supplied so much useful information. We wish to acknowledge permission given by Mr. R. A. Bowman, M.A., F.R.I.C.; Mr. J. W. Davis, A.R.C.S., B.Sc.; Professor Sir C. K. Ingold, D.Sc., A.R.C.S., F.R.I.C., F.R.S.; Griffin and George Ltd. and John Murray Ltd. to use work from their publications, and to help received from Mr. G. Robinson, F.R.I.C. and Mr. D. H. Tomlin, M.A.

Thanks are due to Mr. G. Mackay, M.A. and Mr. E. L. Vernon, M.Sc., former colleagues of one of us, P. D. Arculus, at King Edward VII School, Sheffield, who provided the initial stimulus to write this book without being aware that they were doing so at the time!

P. D. Arculus
M. J. English

CONTENTS

INTRODUCTORY NOTES

Appendix A contains additional information for certain experiments, the titles of which have been asterisked in the text.

The items required for the individual performance of each experiment are given in the appropriate requirement list: quantities of chemicals are about 20 per cent in excess of the expected minimum. Where material is to be employed communally in any experiment, this is specifically stated. These requirement lists do not include a barometer and wall thermometer for ascertaining laboratory conditions, nor do they include Bunsen burners, stands and the usual pieces of ironmongery, all of which are taken to be readily available. It is also assumed that residue bottles, suitably labelled, are to hand for the collection and ultimate recovery of organic solvents.

A statement such as 'c. 1·0 M' means that the molarity lies within 10 per cent of the stated figure, and likewise for all approximate strengths, volumes and weights of chemicals. As an outcome of this, ignition tubes (75 mm × 10 mm), test tubes (125 mm × 16 mm) and boiling tubes (150 mm × 25 mm) have been freely employed for the measurement of approximate volumes of solutions and liquids and the weights of powders. For the measurement of approximate volumes of expensive solutions, e.g. aqueous potassium iodide, the use of automatic dispensing pipettes is advised, although these have not been incorporated in the requirement list. No definite volumes have been given for indicators—these are best delivered from small plastic dropping bottles.

'Specified volumetric ware' is referred to throughout the book and consists of one of each of the following:

50 cm^3 burette
20 cm^3 pipette
250 cm^3 graduated flask
250 cm^3 conical flask
100 cm^3 beaker
250 cm^3 beakers
c. 20 cm long glass stirring rod
c. 7 cm filter funnel (preferably plastic)
c. 11 cm × 11 cm white tile (preferably plastic)
pipette filler (obtainable from Griffin and George Ltd., Frederick Street, Birmingham 1) or, as an alternative, a suitably sized plastic syringe (from Willen Bros. Ltd., New Cavendish Street, London W.1)

1

REACTION INVESTIGATIONS

Introduction

The establishment of chemical formulae and equations is based upon experimental observations. The accepted formula of any compound can only be altered as the outcome of new experimental evidence. The chemical change between certain reactants, however, may vary according to the conditions and thus cannot be represented by a single equation.

The reaction between magnesium and hydrochloric acid of any concentration is

$$Mg + 2HCl \longrightarrow MgCl_2 + H_2$$

i.e.

$$Mg + 2H^+ \longrightarrow Mg^{2+} + H_2.$$

This same ionic equation holds for the reaction between magnesium and very dilute nitric acid. With an increasing concentration of nitric acid, the course of the principal reaction varies profoundly: with bench dilute acid it is

$$3Mg + 8HNO_3 \longrightarrow 3Mg(NO_3)_2 + 2NO + 4H_2O$$

i.e.

$$3Mg + 8H^+ + 2NO_3^- \longrightarrow 3Mg^{2+} + 2NO + 4H_2O$$

and with concentrated nitric acid it is

$$Mg + 4HNO_3 \longrightarrow Mg(NO_3)_2 + 2NO_2 + 2H_2O$$

i.e.

$$Mg + 4H^+ + 2NO_3^- \longrightarrow Mg^{2+} + 2NO_2 + 2H_2O.$$

An equation can only represent a chemical change which has taken place and can never predict what will happen in other conditions.

Qualitative experiments may indicate the nature of the products of a reaction, but quantitative investigations are essential for the formulation of the correct stoichiometric equation. An established equation is a qualitative commentary on the reaction it represents and, furthermore, it provides a quantitative statement on the ratio of the number of moles of reactants and products concerned.

The chemical equations already given above are not ideally informative about the nature and conditions of chemical changes. Further explanation is required as to the physical state of the species involved,

e.g. $Mg(s) + 4H^+(aq) + 2NO_3^-(aq) \longrightarrow Mg^{2+}(aq) + 2NO_2(g) + 2H_2O(l)$

Solid (*s*), liquid (*l*), gas (*g*) and aqueous solution (*aq*) are notations which are used throughout the book.

Even this informative ionic equation is inadequate as far as energy exchanges are concerned!

1.1 Investigations into the validity of an experimental method using the reaction between sodium and water*

Theory

When sodium reacts with water a volume of hydrogen, dependent on the mass of the sodium, is displaced, and this volume may be determined by subsequent collection and measurement. After the reaction has ceased, the sodium hydroxide solution produced, which is also dependent on the mass of the sodium, is estimated by titration with a standard solution of hydrochloric acid.

A comparison of the two sets of results should lead to a discussion of the adequacy of the experimental techniques employed.

Requirements

1·5 g sodium (CARE)
200 cm³ 0·1M hydrochloric acid
120 cm³ 10:1 by volume industrial alcohol/water solution
methyl red indicator solution
specified volumetric ware
250 cm³ measuring cylinder
three *dry* ignition tubes
−10°C to 110°C thermometer
cork borer with boring to fit an ignition tube
3 cm long cork
Winchester 'quart' bottle fitted with delivery tubes, a 250 cm³ conical flask, a Mohr spring clip and a c. 700 cm³ bottle, assembled as shown:

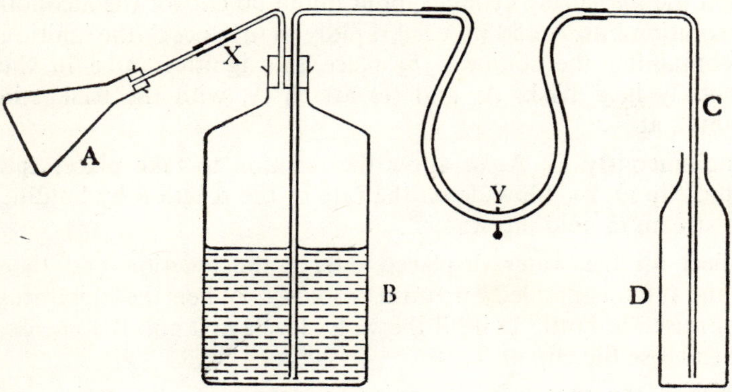

FIG. 1

Procedure

1. Assemble the Winchester 'quart' displacement apparatus, as shown, but disconnect the reaction flask A. Blow, at X, into B to fill the syphon tube C with water and then proceed to test the apparatus for leaks by (*a*) pinching the rubber delivery tube at X, when the water should stop syphoning, and (*b*) attaching A (as in the diagram) at X, when the flow of water should again stop. If the water continues to syphon, there is a leak which must be checked.

2. Again assemble the apparatus with A disconnected and B at least half-full of water. Make the water syphon through C and then apply the clip at Y to stop the water running.

3. Using the measuring cylinder, put 100 cm^3 of water in D and place the end of C under the surface of the water.

4. Bore a hole in the cork so that the boring tightly fits into the ignition tubes. Retain a short length of the boring to use as a cork for the ignition tubes.

5. Fit a *dry* ignition tube with the cork boring, support it in the bored cork and weigh.

6. Away from water and with care, cut a piece of bright sodium, of bulk about that of a cigarette filter tip (i.e. c. 0·5 g), rapidly shave it into the ignition tube (to fill it with lightly packed shavings) and immediately cork. Weigh this ignition tube, again supported in the bored cork.

7. From the measuring cylinder, pour about 50 cm^3 of the alcohol/water solution into A and proceed rapidly to (*a*) uncork the ignition tube containing the sodium, (*b*) place this ignition tube in the horizontally held flask, A, and (*c*) attach A, with the flask still horizontal, at X.

8. Simultaneously tilt A, to allow the reaction to take place, and open the clip at Y. Slow down the rate of the reaction by holding A in a stream of cold tap water.

9. Collect all the water displaced during the reaction and then allow the fully assembled apparatus to cool. When the apparatus is cold, raise the bottle D until the water levels in it and B coincide, and then close the clip at Y.

10. Employ the measuring cylinder to determine the volume of water displaced by the reaction.

11. Record the laboratory temperature and barometric pressure, and look up and record the S.V.P. of water at the laboratory temperature.

12. Transfer the reaction solution and washings to a 250 cm^3 graduated flask, make up to the mark with purified water and shake well. Titrate 20 cm^3 portions of this alkaline solution with the standard 0·1M hydrochloric acid, using methyl red as indicator.

13. Repeat steps *2*, *3*, and *5* to *12* once again, ensuring that the bung is inserted to the same extent in step *7* (*c*).

Observations and Calculations

1. Laboratory temperature = °C
Barometric pressure = cm Hg
S.V.P. of water at the laboratory temperature = cm Hg
Density of hydrogen at S.T.P. = 0·000089 g/cm^3

2. Complete a table with the following headings:

g of sodium sample	cm^3 of hydrogen displaced under laboratory conditions	cm^3 of hydrogen at S.T.P.	g of hydrogen displaced	g of sodium based on g of hydrogen displaced	cm^3 of 0·1M acid to neutralise 20 cm^3 alkali produced	cm^3 of 0·1M acid to neutralise total alkali produced	g of sodium calculated from total alkali produced

Questions

1. What comments can be made on the results obtained in the fifth and eighth columns above?

2. Why is an alcohol/water solution employed in the reaction with sodium?

3. What is the equation for the reaction between ethyl alcohol and sodium? What is the name of the unusual product? What is the equation for the action of water on this unusual product?

4. Why is it necessary to equalise the water levels in step *9*?

5. What possible sources of error are present in the experiment?

6. What improvements might be made to the experimental techniques?

.2 Establishment of the ratio of the number of moles of reactants for the reaction between persulphate and iron(II) ions in acid solution

Theory

Potassium permanganate acts as an oxidising agent in acid solution:

$$MnO_4^-(aq)+8H^+(aq)+5e \longrightarrow Mn^{2+}(aq)+4H_2O(l)$$

and will quantitatively oxidise iron(II) ions to iron(III) ions:

$$Fe^{2+}(aq) \longrightarrow Fe^{3+}(aq)+e.$$

Hence, permanganate ions may be employed for the titration of pipetted portions of ammonium iron(II) sulphate solution, the end point being when the pink colour of permanganate ions just persists.

Other oxidising agents, such as persulphate ions, similarly oxidise iron(II) ions in acid solution. A weighed sample of potassium persulphate is added to an excess fixed volume of a standard solution of ammonium iron(II) sulphate, and the iron(II) ions unaffected by the persulphate ions are estimated using a standardised solution of potassium permanganate.

Requirements

0·6 g A.R. potassium persulphate
280 cm^3 0·1M ammonium iron(II) sulphate
200 cm^3 c. 0·02M potassium permanganate solution
100 cm^3 bench dilute sulphuric acid
specified volumetric ware plus a 50 cm^3 pipette
test tube
boiling tube

Procedure

1. Pipette 20 cm^3 of the standard ammonium iron(II) sulphate solution into a conical flask, add half a test tube (c. 10 cm^3) of bench dilute sulphuric acid, and titrate with the approximately 0·02M potassium permanganate solution. Repeat the titration to obtain concordant results and thus standardise the potassium permanganate solution.

2. Place a funnel in the neck of a 250 cm^3 graduated flask. Weigh, accurately, a 100 cm^3 beaker containing about 0·5 g of the potassium persulphate, empty the contents into the funnel and reweigh the beaker (i.e. weighing by difference).

3. Wash the oxidising agent through into the graduated flask, and in the process use *no more* than one boiling tube (c. 55 cm^3) of purified water.

4. By using three 50 cm^3 pipetted portions, add exactly 150 cm^3 of ammonium iron(II) sulphate solution to the contents of the graduated flask. Make up to the mark with purified water and shake well.

5. Allow the contents of the flask to stand for at least five minutes and reshake before use. Titrate 40 cm^3 (two 20 cm^3 pipetted portions) of the mixture, in the presence of half a test tube (c. 10 cm^3) of bench dilute sulphuric acid, with the standardised potassium permanganate solution, to obtain concordant results.

Calculations

1. For each permanganate-iron(II) titration, determine:

(*a*) the number of moles of ammonium iron(II) sulphate,
(*b*) the number of moles of potassium permanganate,
(*c*) the molarity of the potassium permanganate solution.

2. For each permanganate-persulphate/ammonium iron(II) titration, determine:

(*a*) the number of moles of persulphate ions,

(*b*) the initial number of moles of ammonium iron(II) sulphate in the pipetted portion,

(*c*) the number of moles of potassium permanganate in the titre and thus the number of moles of ammonium iron(II) sulphate equivalent to it,

(*d*) the number of moles of ammonium iron(II) sulphate oxidised by persulphate ions,

(*e*) the ratio of the number of moles of ammonium iron(II) sulphate to persulphate ions.

Questions

1. What is the ionic equation for the reaction between permanganate and iron(II) ions in acid solution?

2. The ammonium iron(II) sulphate solution is made up in dilute sulphuric acid and a further volume is added for the titration. Why cannot hydrochloric acid or nitric acid be employed?

3. What is the ionic equation for the reaction between persulphate and iron(II) ions?

4. What is the half ionic equation for the reduction of persulphate ions?

5. What are the ionic equations for the oxidation by persulphate ions, in acid solution, of (*a*) sulphite to sulphate ions and (*b*) lead(II) ions to lead(IV) oxide?

1.3 Determination of the composition of a given acid oxalate, $K_xH_y(C_2O_4)_z$*

Theory

Oxalate ions can act as a reducing agent:

$$\begin{matrix} COO^- \\ | \\ COO^- \end{matrix}(aq) \longrightarrow 2CO_2(g)+2e$$

and will quantitatively reduce permanganate ions to manganese(II) ions:

$$MnO_4^-(aq)+8H^+(aq)+5e \longrightarrow Mn^{2+}(aq)+4H_2O(l).$$

This oxidation-reduction reaction may be employed for the titration of pipetted portions of sodium oxalate with potassium permanganate in the presence of dilute sulphuric acid at about 70°C.

The hydrogen ions of the acid oxalate may be estimated by a normal acid-base titration.

Requirements

2·0 g A.R. sodium oxalate
200 cm³ acid oxalate solution containing 12·8 g/1
100 cm³ 0·1M sodium hydroxide solution
200 cm³ c. 0·04M potassium permanganate solution
200 cm³ bench dilute sulphuric acid
phenolphthalein indicator solution
specified volumetric ware plus a 100 cm³ graduated flask
test tube
boiling tube

Procedure

1. Standardisation of the potassium permanganate solution

(*a*) Weigh out accurately about 1·4 g of the sodium oxalate into a 100 cm^3 beaker, add a boiling tube (c. 55 cm^3) of purified water and gently warm, with stirring, to dissolve the solid. Transfer the solution and washings to the 100 cm^3 graduated flask, make up to the mark with purified water and shake well to produce a standard solution.

(*b*) Pipette 20 cm^3 of this standard solution into a conical flask, add a test tube (c. 20 cm^3) of bench dilute sulphuric acid, and heat the contents of the flask until bubbles start to rise from the solution (c. 70°C). Titrate this hot solution with the potassium permanganate solution until the pink colour of the permanganate ions just persists. (If the solution turns brown, insufficient sulphuric acid has been added, and this must be rectified by addition of a further volume.)

2. Estimation of the acid oxalate as an acid

Titrate pipetted 20 cm^3 portions of the acid oxalate with 0·1M sodium hydroxide solution, using phenolphthalein as indicator, to obtain concordant results.

3. Estimation of the acid oxalate as a reducing agent

Repeat step *1*(*b*) above, employing the acid oxalate solution provided instead of the sodium oxalate.

Calculations

1. For the standardisation, determine:

(*a*) the molarity of the sodium oxalate,

(*b*) for the titration, the number of reacted moles of (*i*) sodium oxalate and (*ii*) potassium permanganate,

(*c*) the molarity of the potassium permanganate.

2. For the acid oxalate as an acid, determine the molarity of the hydrogen ions.

3. For the acid oxalate as a reducing agent, determine:

(*a*) for the titration, the number of reacted moles of (*i*) potassium permanganate and (*ii*) oxalate ions,

(*b*) the molarity of the oxalate ions.

4. For the acid oxalate,

(*a*) deduce the molarity of the potassium ions per litre,

(*b*) elicit the simplest ratio x:y:z,

(*c*) record its empirical formula.

Questions

1. Why must potassium permanganate be standardised prior to use and not made up directly into a solution of known molarity?

2. What is the ionic equation for the reaction between oxalate and permanganate ions in acid solution?

3. Why is it necessary to heat the oxalate solution to about 70°C before immediate titration with potassium permanganate solution, and why should further heating be unnecessary?

4. Employing the empirical formula of the acid oxalate, what are the equations for the reactions between the acid oxalate and (*a*) sodium hydroxide, (*b*) potassium permanganate in acid conditions?

.4 Determination of the equation for the reaction between iron(II) oxalate and potassium permanganate in acid solution*

Requirements

5·0 g ammonium iron(II) sulphate crystals
100 cm^3 0·03M iron(II) oxalate in dilute sulphuric acid
200 cm^3 c. 0·02 potassium permanganate solution
200 cm^3 bench dilute sulphuric acid
specified volumetric ware plus a 100 cm^3 graduated flask
test tube
boiling tube

Procedure

1. Weigh out accurately about 3·9 g of ammonium iron(II) sulphate crystals into a 100 cm^3 beaker.

2. Dissolve these crystals in a boiling tube (c. 55 cm^3) of bench dilute sulphuric acid, and transfer the solution and washings to the 100 cm^3 graduated flask. Make up to the mark with bench dilute sulphuric acid and shake well.

3. Pipette 20 cm^3 of the standard ammonium iron(II) sulphate solution into a conical flask, add half a test tube (c. 10 cm^3) of bench dilute sulphuric acid, and titrate with the potassium permanganate solution until the pink colour of the permanganate ions just persists.

4. Repeat step *3* to obtain concordant results.

5. Pipette 20 cm^3 of the standard iron(II) oxalate solution into a conical flask, add half a test tube of bench dilute sulphuric acid, and heat the contents of the flask until bubbles start to rise from the solution (c. 70°C).

6. Titrate this hot solution with the standardised potassium permanganate solution.

7. Repeat steps *5* and *6* to obtain concordant results.

Calculations

Determine:

(*a*) the molarity of the ammonium iron(II) sulphate solution,
(*b*) the molarity of the potassium permanganate solution,
(*c*) the number of moles of iron(II) oxalate used in each titration,
(*d*) the number of moles of potassium permanganate employed in each titration,
(*e*) the ratio of the number of moles for the reaction between iron(II) oxalate and potassium permanganate in acid solution.

Questions

1. What is the equation for the reaction under consideration?

2. (*a*) What happens when dichromate ions are added to an acid solution of (*i*) iron(II) ions and (*ii*) oxalate ions? Investigate qualitatively.

(*b*) In the light of these test experiments, what is the equation for the reaction between iron(II) oxalate and potassium dichromate in acid solution?†

1.5 Investigation of the reaction between iron(II) oxalate and potassium chlorate in acid solution*

Requirements

3·0 g potassium chlorate
200 cm^3 solution of iron(II) oxalate in dilute sulphuric acid
160 cm^3 0·02M potassium permanganate solution
250 cm^3 bench dilute sulphuric acid
specified volumetric ware plus a 500 cm^3 conical flask
test tube

† This may be quantitatively investigated in the manner already outlined, by employing potassium dichromate solution in place of potassium permanganate solution.

Procedure

1. Slowly titrate a 20 cm^3 portion of the iron(II) oxalate solution with 0·02M potassium permanganate solution under the normal oxalate conditions.

2. Repeat the titration to obtain concordant titres.

3. Weigh about 2·5 g of potassium chlorate and dissolve in approximately 250 cm^3 of water contained in the 500 cm^3 conical flask.

4. To a 20 cm^3 portion of the iron(II) oxalate solution add two test tubes (c. 40 cm^3) of the potassium chlorate solution and about an equal volume of bench dilute sulphuric acid. After about one minute, boil the mixture gently for approximately three minutes.† Slowly titrate the hot mixture with the 0·02M potassium permanganate solution.

5. Repeat step *4* to obtain coincident volumes of permanganate.

6. Qualitatively investigate the products of the reaction between iron(II) oxalate and potassium chlorate in acid solution.

Calculations

1. Calculate the concentration of anhydrous iron(II) oxalate in grams per litre.

2. Determine the ratio of the number of moles of iron(II) oxalate and potassium chlorate for their reaction in acid conditions.

Question

What is the equation for the reaction under consideration?

1.6 Titrimetric investigations into the reaction between formic acid and potassium permanganate in aqueous solution*

Requirements

225 cm^3 formic acid solution
250 cm^3 0·05M sodium oxalate solution
175 cm^3 0·1M sodium hydroxide solution
400 cm^3 c. 0·02M potassium permanganate solution
400 cm^3 bench dilute sulphuric acid
250 cm^3 bench sodium carbonate solution
phenolphthalein indicator solution
specified volumetric ware plus a 25 cm^3 pipette
test tube

† Under these conditions the chlorate exerts its full oxidising effect on the iron(II) oxalate.

Procedure

1. Standardise the potassium permanganate solution by titrating 25 cm^3 pipetted portions of the standard sodium oxalate solution with it, under the normal oxalate conditions.

2. Standardise the formic acid solution by titrating 20 cm^3 pipetted portions with the standard sodium hydroxide solution, employing phenolphthalein as indicator.

3. Pipette 20 cm^3 of the formic acid solution into the conical flask, and add half a test tube (c. 10 cm^3) of bench dilute sulphuric acid. Titrate with the standardised potassium permanganate solution until a pink colour persists, then boil and, if necessary, add more titrate. There is no need to repeat this titration.

4. (*a*) Pipette 20 cm^3 of the formic acid solution into the conical flask, add half a test tube (c. 10 cm^3) of bench sodium carbonate solution, and heat until bubbles start to rise from the solution (i.e. c. 70°C). Periodically reheat the solution while titrating with the standardised potassium permanganate solution. Detect the end point by holding up the conical flask towards a good light and looking through the edge of its contents.

(*b*) Add a further 5·0 cm^3 of the standardised solution of potassium permanganate to obtain a definite pink colour.

(*c*) Add a 25 cm^3 pipetted portion of the standard sodium oxalate solution (i.e. in excess). Carefully add dilute sulphuric acid until effervescence ceases and follow with a half test tube of acid (c. 10 cm^3) to ensure adequate acidity. Heat, almost to boiling, until the precipitate has disappeared.

(*d*) Immediately titrate the hot solution with the standardised potassium permanganate solution.

(*e*) Repeat steps *4* (*a*) to (*d*) to obtain concordant values.

Question

From calculations on the ratios of the number of moles of reactants involved, what interpretation may be given to the results of this titrimetric investigation?

1.7 An experimental sequence involving iodate and iodide ions and iodine

A Establishment of the equation for the reaction between iodate and iodide ions in the presence of approximately 0·5M hydrochloric acid

Requirements

150 cm^3 0·1M sodium thiosulphate solution
100 cm^3 0·025M potassium iodate solution
50 cm^3 c. 10 per cent potassium iodide solution
50 cm^3 bench dilute hydrochloric acid
starch indicator
specified volumetric ware
test tube

Procedure

1. Pipette 20 cm^3 of the 0·025M potassium iodate solution into a conical flask, and add half a test tube (c. 10 cm^3) of the potassium iodide solution and an equal volume of bench dilute hydrochloric acid.

2. Titrate the liberated iodine with the standard 0·1M sodium thiosulphate solution; when the iodine solution reaches a pale straw colour, starch may be used as indicator.

3. Repeat steps *1* and *2* to obtain concordant titres.

Questions

1. In the reaction under investigation, how many moles of (*a*) potassium iodate were used and (*b*) iodine were produced, at each titration?

2. What is (*a*) the molecular equation and (*b*) the ionic equation for the reaction between potassium iodate and potassium iodide in the presence of approximately 0·5M hydrochloric acid to produce iodine?

*B Establishment of the equation for the reaction between iodate ions and iodine in the presence of approximately 5M hydrochloric acid**

Requirements

100 cm^3 0·1M sodium thiosulphate solution
100 cm^3 0·05M potassium iodate solution
100 cm^3 c. 0·1M alcoholic iodine solution

100 cm^3 concentrated hydrochloric acid
10 cm^3 chloroform
starch indicator
specified volumetric ware plus a 10 cm^3 pipette
250 cm^3 glass-stoppered reagent bottle
ignition tube
test tube

Procedure

1. Standardise the iodine solution by titrating 10 cm^3 pipetted portions of it with the standard 0·1M sodium thiosulphate solution run in from a burette (starch may be employed as an indicator).

2. (*a*) Pipette 10 cm^3 of the iodine solution into the reagent bottle. Add one test tube (c. 20 cm^3) of concentrated hydrochloric acid and half an ignition tube (c. 2 cm^3) of chloroform.

(*b*) From the burette, run in the 0·05M potassium iodate solution; after each addition, stopper and vigorously shake the reagent bottle. The end point, which is very sharp, is reached when the pink colour of the iodine in the chloroform disappears and the aqueous layer becomes pale yellow owing to the formation of iodine monochloride.

3. Repeat the titration procedure of step *2* to obtain concordant results.

Questions

1. Why is an alcoholic solution of iodine employed?

2. How many moles of potassium iodate and iodine react with each other in the titration?

3. What is (*a*) the molecular equation and (*b*) the ionic equation for the reaction between potassium iodate and iodine in the presence of approximately 5M hydrochloric acid?

4. (*a*) What are the physical and chemical characteristics of iodine monochloride, and what is peculiar about its appearance here as a titration product?

(*b*) What are the physical and chemical characteristics of other iodine interhalogen compounds?

C *Establishment of the equation for the reaction between iodate and iodide ions in the presence of approximately 5M hydrochloric acid*

Requirements

40 cm^3 0·05M potassium iodate solution
50 cm^3 0·1M potassium iodide solution
100 cm^3 concentrated hydrochloric acid
10 cm^3 chloroform
specified volumetric ware plus a 10 cm^3 pipette
250 cm^3 glass-stoppered reagent bottle
ignition tube
test tube

Procedure

1. (*a*) Pipette 10 cm^3 of the 0·1M potassium iodide solution into the reagent bottle. Add one test tube (c. 20 cm^3) of concentrated hydrochloric acid, followed by half an ignition tube (c. 2 cm^3) of chloroform.

(*b*) Titrate with the 0·05M potassium iodate solution; stopper the bottle and shake its contents vigorously after each addition. The end point occurs when the chloroform layer, which turns pink owing to iodine formation on the first addition of iodate solution, becomes colourless and the pale yellow colour of iodine monochloride is seen in the aqueous layer.

2. Repeat step *1* to obtain concordant results.

Questions

1. What explanation can be given for the appearance of iodine upon the first addition of iodate solution?

2. How many moles of potassium iodate and potassium iodide react together in the titration?

3. What is (*a*) the molecular equation and (*b*) the ionic equation for the reaction between potassium iodate and potassium iodide in the presence of approximately 5M hydrochloric acid?

4. How may this reaction equation be derived by a consideration of the equations experimentally established in A and B of this sequence?

*D Determination of the formula of an acid iodate of potassium $KIO_3.xHIO_3$**

Requirements

4·0 g acid iodate
100 cm^3 0·05M sodium hydroxide solution
50 cm^3 0·2M potassium iodide solution
100 cm^3 concentrated hydrochloric acid
10 cm^3 chloroform
methyl red indicator solution
specified volumetric ware plus a 10 cm^3 pipette
250 cm^3 glass-stoppered reagent bottle
ignition tube
test tube

Procedure

1. Weigh accurately about 3·25 g of the acid iodate into a 100 cm^3 beaker, add a boiling tube (c. 55 cm^3) of purified water, warm, and stir to obtain a solution. Transfer the solution and washings to a 250 cm^3 graduated flask, make up to the mark with purified water, and shake well.

2. Estimate the acid iodate as an acid by titrating 20 cm^3 pipetted portions of the 0·05M sodium hydroxide solution with the acid iodate solution, employing methyl red as indicator.

3. Estimate the acid iodate as an iodate by titration with the 0·2M potassium iodide solution in the presence of approximately 5M hydrochloric acid as follows:

(*a*) Pipette 10 cm^3 of the iodide solution into the reagent bottle, and add one test tube (c. 20 cm^3) of concentrated hydrochloric acid and half an ignition tube (c. 2 cm^3) of chloroform.

(*b*) From the burette, run in the iodate solution; after each addition, stopper and vigorously shake the reagent bottle. The end point, which is very sharp, is reached when the pink colour of the iodine in the chloroform disappears.

(*c*) Repeat steps (*a*) and (*b*) to obtain coincident titres.

Calculations

1. For the acid iodate solution, (*a*) calculate the molarity of hydrogen ions and (*b*) by reference to the equation established in C of this sequence, calculate the molarity of the iodate ions.

2. For the acid iodate molecule, state (*a*) the ratio of the number of moles of hydrogen ions to iodate ions and (*b*) the formula.

Question

By what alternative means to that employed might the acid iodate have been estimated as an iodate?

E Estimation of the iodine and iodide ion contents of an aqueous solution*

Theory

1. The iodine content is estimated by direct titration with standard sodium thiosulphate solution:

$$I_2(aq)+2Na_2S_2O_3(aq) \longrightarrow NaI(aq)+2Na_2S_4O_6(aq).$$

2. The iodine and the iodide ion contents are estimated by direct titration, in the presence of approximately 5M hydrochloric acid, with standard potassium iodate solution:

$$2I_2(aq)+KIO_3(aq)+6HCl(aq) \longrightarrow 5ICl(aq)+KCl(aq)+3H_2O(l)$$
$$2KI(aq)+KIO_3(aq)+6HCl(aq) \longrightarrow 3ICl(aq)+3KCl(aq)+3H_2O(l).$$

Requirements

150 cm³ of an aqueous solution containing iodine and iodide ions
70 cm³ 0·05M potassium iodate solution
60 cm³ 0·1M sodium thiosulphate solution
100 cm³ concentrated hydrochloric acid
10 cm³ chloroform
starch indicator
specified volumetric ware plus a 10 cm³ pipette
250 cm³ glass-stoppered reagent bottle
ignition tube
test tube

Procedure

1. Estimate the iodine content of the solution by titration of 20 cm³ pipetted portions of it with the standard 0·1M sodium thiosulphate solution (starch may be employed as an indicator).

2. (*a*) Pipette 10 cm^3 of the iodine/iodide solution into the reagent bottle. Add one test tube (c. 20 cm^3) of concentrated hydrochloric acid and half an ignition tube (c. 2 cm^3) of chloroform.

(*b*) Run in the 0·05M potassium iodate solution from a burette, stoppering the reagent bottle and shaking after each addition, until the colour of the chloroform layer is discharged.

3. Repeat step *2* to obtain concordant titration results.

Calculations

Determine:

(*a*) the concentration of iodine expressed in (*i*) moles per litre and (*ii*) grams per litre, in the iodine/iodide solution,

(*b*) the volume of 0·05M potassium iodate solution which reacts at each titration with (*i*) the iodine and (*ii*) the iodide, in 20 cm^3 of the iodine/iodide solution,

(*c*) the concentration of potassium iodide, expressed in (*i*) moles per litre and (*ii*) grams per litre, in the iodine/iodide solution.

Question

In what other way could the iodine and iodide ion contents of the given solution be estimated?

1.8 Investigation of the reaction between copper(II) and ammonium ions in aqueous solution*

Theory

Copper(II) ions in slightly acid solution react with iodide ions as follows:

$$2Cu^{2+}(aq)+4I^-(aq) \longrightarrow 2CuI(s)+I_2(aq).$$

With the addition of excess iodide ions, the liberation of iodine is quantitative with respect to copper(II) ions. The precipitate of copper(I) iodide is off-white in colour and in no way interferes with the titration of the liberated iodine with standard sodium thiosulphate solution. By this means, the reaction between copper(II) sulphate and ammonia, in aqueous solution, may be investigated.

Requirements

120 cm³ 0·1M copper(II) sulphate solution
120 cm³ 0·2M ammonium hydroxide solution
120 cm³ 0·1M sodium thiosulphate solution
120 cm³ 10 per cent potassium iodide solution
bench sodium carbonate solution
bench dilute hydrochloric acid
bench dilute acetic acid
starch indicator
specified volumetric ware plus ten 250 cm³ conical flasks and five filter funnels
five adhesive labels

Procedure

1. Label and number five of the conical flasks from 1 to 5 and make up the contents of each as in the following table:

number of conical flask	cm³ of 0·1M copper(II) sulphate	cm³ of 0·2M ammonium hydroxide	
		either	or
1	20	0	0
2	20	5	10
3	20	15	20
4	20	25	30
5	20	35	40

2. Take each mixture in turn and filter off the precipitate through a fluted filter paper. Wash the precipitate with purified water, add the washings to the filtrate, and retain this solution and the washed precipitate for further treatment.

(*a*) *Precipitate*

(*i*) Transfer the precipitate and filter paper to a conical flask and just dissolve the precipitate in dilute hydrochloric acid.

(*ii*) Neutralise any excess acid by adding dilute sodium carbonate solution until a precipitate just forms.

(*iii*) Just destroy this precipitate with a few drops of dilute acetic acid, thereby making the solution slightly acidic.

(*iv*) Estimate the copper originally present in the precipitate by adding excess (half a test tube, that is, c. 10 cm³) of the potassium iodide solution and titrating the liberated iodine with the 0·1M sodium thiosulphate solution, using starch as indicator.

(*b*) *Filtrate and washings of the pale blue colour of hydrated copper(II) ions*

(*i*) Add dilute sodium carbonate solution until a precipitate just forms.

(*ii*) Repeat as for precipitate instructions (*iii*) and (*iv*).

(*c*) *Filtrate and washings of the dark blue colour of ammonia-copper(II) complex ions*

(*i*) Add dilute acetic acid until the blue colour of the complex is just destroyed.

(*ii*) Repeat as for precipitate instructions (*ii*) to (*iv*).

Observations and Calculations

1. Complete a table with the following headings:

number of conical flask	cm^3 of 0·1M copper(II) sulphate	cm^3 of 0·2M ammonium hydroxide	for precipitate		for filtrate	
			cm^3 of 0·1M sodium thiosulphate	% of total copper present	cm^3 of 0·1M sodium thiosulphate	% of total copper present

2. Draw, using the same axes, graphs of (*a*) the percentage of the total copper present in the precipitate and (*b*) the percentage of the total copper present in the filtrate against the volume of ammonium hydroxide solution added.

3. By reference to the graphs, derive equations for the reactions between copper(II) sulphate and ammonium hydroxide in aqueous solution (*a*) for maximum precipitation of copper(II) hydroxide and (*b*) for the production of the ammonia-copper(II) ion, thus leading to the formula of the complex ion.

1.9 Investigation of the reaction between ammonium ions and formaldehyde*

Theory

Formaldehyde solution is standardised by oxidation with excess iodine in alkaline solution:

$$HCHO(aq) + I_2(aq) + H_2O(l) \longrightarrow HCOOH(aq) + 2HI(aq)$$

and the residual iodine is determined by titration with standard sodium thiosulphate solution.

Formaldehyde solution reacts with ammonium ions as follows:

$$6HCHO(aq)+4NH_4^+(aq) \longrightarrow (CH_2)_6N_4(aq)+4H^+(aq)+6H_2O(l).$$

The hydrogen ions produced are estimated by addition of excess standard alkali and back titration with standard acid.

Requirements

210 cm³ formaldehyde solution
250 cm³ 0·05M sodium thiosulphate solution
50 cm³ 1·0M ammonium chloride solution
120 cm³ 1·0M sodium hydroxide solution
100 cm³ 1·0M hydrochloric acid
180 cm³ c. 0·05M iodine solution
20 cm³ bench sodium hydroxide solution
20 cm³ bench dilute hydrochloric acid
starch indicator
bromothymol blue indicator solution
phenolphthalein indicator solution
specified volumetric ware plus a 5 cm³ pipette, a 25 cm³ pipette, a 50 cm³ pipette and a 100 cm³ graduated flask
test tube
spotting rod

Procedure

1. Standardisation of the iodine solution

Titrate 10 cm³ pipetted portions of the iodine solution with the standard sodium thiosulphate solution to obtain concordant titres (starch may be employed as indicator when the solution has paled to a straw colour).

2. Standardisation of the formaldehyde solution

(*a*) Pipette 50 cm³ of the 0·05M sodium thiosulphate solution into a 250 cm³ graduated flask. Dilute to the mark with purified water and shake well to obtain a 0·01M sodium thiosulphate solution.

(*b*) (*i*) Attach the pipette filler to the 5 cm³ pipette and transfer a 5 cm³ portion of the formaldehyde solution to a 250 cm³ graduated flask.

(*ii*) Pipette 100 cm³ (two 50 cm³ portions) of the standardised iodine solution into the graduated flask and follow by the slow addition, with constant swirling, of sufficient bench sodium hydroxide solution (about a test tube of it) to obtain a permanent yellow colour.

(*iii*) Allow the contents of the graduated flask to stand for fifteen minutes.

(*iv*) Acidify with bench dilute hydrochloric acid; test for achievement of acidity by spotting the solution onto blue litmus paper (remove the masking iodine colour by dropping the paper into a test tube of the diluted 0·01M sodium thiosulphate solution).

(*v*) Make up to the mark with purified water and titrate 50 cm^3 pipetted portions of this solution as in step *1*, using the 0·1M sodium thiosulphate solution as titrant.

3. Investigation of the reaction

(*a*) With excess ammonium ions

(*i*) In the 100 cm^3 graduated flask, prepare, by pipette, a mixture of 20 cm^3 of the formaldehyde solution, 20 cm^3 of the 1·0M ammonium chloride solution and 50 cm^3 of the 1·0M sodium hydroxide solution, and then make up to the mark with purified water. Shake and leave to stand for fifteen minutes.

(*ii*) Titrate 20 cm^3 pipetted portions of the prepared solution with the 1·0M hydrochloric acid, using bromothymol blue as indicator to obtain coincident results.

(*b*) With excess formaldehyde

(*i*) In a 250 cm^3 graduated flask, prepare, by pipette, a mixture of 150 cm^3 of the formaldehyde solution, 20 cm^3 of the 1·0M ammonium chloride solution and 50 cm^3 of the 1·0M sodium hydroxide solution, and then make up to the mark with purified water. Shake and leave to stand for five minutes.

(*ii*) In the meantime, pipette 25 cm^3 of the 1·0M hydrochloric acid into a 250 cm^3 graduated flask. Make up to the mark with purified water and shake to give 0·1M hydrochloric acid.

(*iii*) Titrate 20 cm^3 pipetted portions of the prepared solution with the 0·1M hydrochloric acid, using phenolphthalein as indicator to obtain coincident results.

Calculations

1. Enumerate the molarities of (*a*) the iodine solution and (*b*) the formaldehyde solution.

2. Ascertain the ratio of the number of males of (*a*) formaldehyde taken to hydrogen ions produced in the presence of excess ammonium ions and (*b*) ammonium ions taken to hydrogen ions produced in the presence of excess formaldehyde.

Questions

1. In the light of the given reaction equation, what comments may be made about the results of calculations *2* (*a*) and (*b*)?

2. For $(CH_2)_6N_4$, (*a*) what are its systematic and alternative names, (*b*) what is its probable structural formula, (*c*) what are its physical characteristics and (*d*) what are its uses and what is manufactured from it?

3. What are the equations for the reaction of ammonia with (*a*) acetaldehyde, (*b*) acetone and (*c*) benzaldehyde?

1.10 Investigations into the reaction between bromine water and sodium thiosulphate solution*

Requirements

325 cm³ bromine water
375 cm³ 0·1M sodium thiosulphate solution
200 cm³ c. 0·05M iodine solution
170 cm³ 10 per cent potassium iodide solution
starch indicator
potassium iodide/starch paper
specified volumetric ware plus a 50 cm³ burette
100 cm³ measuring cylinder
spotting rod
reagents and apparatus for qualitative analysis

Procedure

1. Titrate 20 cm³ portions of the iodine solution with the 0·1M sodium thiosulphate solution (when the iodine solution has reached a pale straw colour, starch may be used as indicator). Obtain coincident titres.

2. To 20 cm³ of the bromine water, measured by means of a burette, add three-quarters of a test tube (c. 15 cm³) of the potassium iodide solution. Titrate the liberated iodine with the 0·1M sodium thiosulphate solution (starch may be used again as indicator).

3. (*a*) Directly titrate 20 cm³ portions of the bromine water, delivered from a burette, with the 0·1M sodium thiosulphate solution. The end point may be found by taking one drop of the reaction mixture and placing it on a piece of potassium iodide/starch paper (when bromine is in excess, a blue colour develops owing to the displacement of iodine by the bromine).

(*b*) Qualitatively examine for the products of the reaction between sodium thiosulphate solution and excess bromine water.

4. (*a*) From the measuring cylinder pour 80 cm^3 of bromine water into a 100 cm^3 graduated flask and, from a burette, add exactly 6 cm^3 of the 0·1M sodium thiosulphate solution. Make the mixture up to the mark by the addition of bromine water. Dispense successive 20 cm^3 portions of this mixture from a burette and titrate each with the 0·1M sodium thiosulphate solution in the manner of step *2*.

(*b*) Qualitatively investigate the nature of the products of the reaction between sodium thiosulphate solution and excess bromine water.

5. (*a*) From the measuring cylinder, pour about 80 cm^3 of the 0·1M sodium thiosulphate solution into a 100 cm^3 graduated flask and, from a burette, add exactly 10 cm^3 of bromine water. Make the mixture up to the mark by the addition of more sodium thiosulphate solution. Titrate 20 cm^3 portions of the iodine solution with this mixture as in step *1*.

(*b*) Qualitatively test for the products of the reaction between bromine water and excess sodium thiosulphate solution.

Calculations

Determine:

(*a*) the molarity of the iodine solution,
(*b*) the molarity of the bromine solution,
(*c*) the equation for the reaction between sodium thiosulphate and excess bromine water,
(*d*) the equation for the reaction between bromine water and excess sodium thiosulphate solution.

1.11 Establishment of the equation for the bromination of phenol*

Theory

Bromate and bromide ions in acid solutions quantitatively liberate bromine:

$$BrO_3^-(aq) + 5Br^-(aq) + 6H^+(aq) \longrightarrow 3Br_2(aq) + 3H_2O(l).$$

Thus a standard solution of potassium bromate in the presence of excess potassium bromide acts as a potential standard bromine solution, which is employed in excess for the direct bromination of phenol.

The residual bromine is estimated by liberation, from iodide ions, of an equimole portion of iodine, followed by the subsequent titration of the iodine with standard sodium thiosulphate solution.

Requirements

3·0 g A.R. potassium bromate
15 g potassium bromide
120 cm^3 phenol solution containing 1·0 g A.R. phenol per litre
150 cm^3 0·1M sodium thiosulphate solution
50 cm^3 10 per cent potassium iodide solution
50 cm^3 concentrated hydrochloric acid
starch indicator
specified volumetric ware plus a 50 cm^3 burette, a 50 cm^3 pipette and a 1000 cm^3 graduated flask
250 cm^3 reagent bottle with a bung to fit
test tube

Procedure

1. Weigh accurately in a 100 cm^3 beaker, about 2·8 g of the A.R. potassium bromate. Add about 12·0 g of potassium bromide and dissolve the solids in the minimum volume of purified water. Transfer the solution and washings to the 1000 cm^3 graduated flask and make up to the mark with purified water.

2. From a burette, dispense 25 cm^3 of the phenol solution into the reagent bottle. Add a 50 cm^3 pipetted portion of the standard bromate-bromide solution and half a test tube (c. 10 cm^3) of concentrated hydrochloric acid. Stopper the reagent bottle, shake and allow to stand for fifteen minutes.

3. Add half a test tube (c. 10 cm^3) of the potassium iodide solution. Titrate the liberated iodine with the 0·1M sodium thiosulphate solution; starch may be employed as indicator when the solution has reached a pale straw colour.

4. Repeat steps *2* and *3* to obtain coincident titres.

Calculations

1. Enumerate the molarities of (*a*) the phenol solution and (*b*) the bromate-bromide solution with respect to potential bromine.

2. Ascertain the ratio of the number of moles of reactants for the bromination of phenol.

Questions

1. What is the equation for the bromination of phenol?

2. What is the systematic name for the white precipitate produced in the reaction?

1.12 Investigation of the precipitation of basic copper(II) compounds*

Requirements

300 cm^3 0·04M copper(II) sulphate solution
100 cm^3 0·1M hydrochloric acid
120 cm^3 0·08M sodium carbonate solution
150 cm^3 0·1M sodium hydroxide solution
methyl orange indicator solution
phenolphthalein indicator solution
specified volumetric ware plus a 50 cm^3 pipette and a 500 cm^3 conical flask
boiling tube

Procedure

1. Investigation of the precipitation of basic copper(II) carbonate by back titration

(*a*) Into the 500 cm^3 conical flask, pipette 50 cm^3 of the copper(II) sulphate solution and bring to the boil.

(*b*) Remove the source of heat and slowly add, by pipette, two 50 cm^3 portions of the sodium carbonate solution.

(*c*) Cautiously boil again and maintain gentle heat for twenty minutes.

(*d*) Cool the solution in a stream of cold tap water until the base of the flask can be held comfortably in the palm of the hand.

(*e*) Filter the liquid through fluted filter paper into a 250 cm^3 graduated flask; wash (the washings must be added to the filtrate) the precipitate on the filter paper with four separate test tube portions (i.e. c. 80 cm^3 in all) of purified water. Make up to the mark with purified water and shake well.

(*f*) Titrate 50 cm^3 pipetted portions of the alkaline filtrate with the standard hydrochloric acid, using methyl orange as indicator to obtain coincident titres.

2. Investigation of the precipitation of basic copper(II) sulphate by direct titration

(*a*) Pipette 50 cm^3 of the copper(II) sulphate solution into a conical flask and add a boiling tube (c. 55 cm^3) of purified water. Heat until almost boiling.

(*b*) Add a quarter of an ignition tube (c. 1 cm^3) of phenolphthalein indicator solution and titrate with the standard sodium hydroxide solution.

(*c*) Repeat steps (*a*) and (*b*) to obtain concordant results.

Calculations

Calculate the reactant mole ratios for the precipitation of (*a*) basic copper(II) carbonate and (*b*) basic copper(II) sulphate.

Questions

1. Why is it possible to titrate copper(II) sulphate solution directly with sodium hydroxide solution?

2. (*a*) What are the possible empirical formulae of the two precipitated compounds? Do they agree with the usually quoted molecular formulae?

(*b*) What are the equations for the precipitation reactions based on these empirical formulae?

3. What are the names and formulae of the naturally occurring basic copper(II) carbonates?

2

SOLUBILITY

Introduction

The principles of equilibrium may be applied to the problem of the solubility of solutes which are appreciably soluble in specified solvents. The equilibria prevailing with sparingly soluble electrolytes, which give rise to ions in aqueous solution, are discussed in the introduction to Chapter 16.

When a solid dissolves in a liquid solvent, the solution process consists of separate particles breaking away from the regular crystal lattice to become part of the disordered liquid phase. This tendency towards maximum disorder or randomness, itself an outcome of disruptive effects of kinetic energy, is responsible for the solution process.

For the equilibrium between a saturated solution and undissolved solid solute, there is a dynamic balance between the rate of solution and the rate at which the solute particles return to the crystal. With many solids, heat is absorbed on dissolution; the crystalline state has a lower potential energy than do the random conditions of solution, and this tendency towards minimum energy favours return of particles to the crystal lattice.

Since the gaseous state is more random than the liquid state, there is an increase in order when gases dissolve in liquids. Gaseous molecules are relatively far apart, but in dissolving they are brought closer together and attract each other more strongly; the potential energy is lowered and heat is evolved.

Solubility data must be determined experimentally, for there is no simple means of predicting whether either increase in randomness or decrease in potential energy is the prevalent factor.

Questions

1. (*a*) What happens when sodium chloride dissolves in water? What energy changes take place?

(*b*) The solubility of sodium chloride in water increases somewhat with a rise in temperature. Why?

2. Why do the solubilities of most gases, in specified solvents, decrease with a rise in temperature?

2.1 Determination of the solubility of ammonium oxalate in water*

Requirements

25 cm^3 saturated solution of ammonium oxalate
100 cm^3 0·02M potassium permanganate solution
100 cm^3 bench dilute sulphuric acid
specified volumetric ware plus a 100 cm^3 graduated flask
test tube

Procedure

1. Pipette 20 cm^3 of the ammonium oxalate solution (take care not to suck up any crystals) into the 100 cm^3 graduated flask, make up to the mark with purified water and shake well.

2. Pipette 20 cm^3 of the diluted ammonium oxalate solution into a conical flask. Add a test tube (c. 20 cm^3) of bench dilute sulphuric acid and heat the contents of the flask until bubbles start to rise from the solution (c. 70°C).

3. Titrate this hot solution with the 0·02M potassium permanganate solution until the pink colour of the permanganate ions just persists. (If the solution turns brown, insufficient sulphuric acid has been added and this must be rectified by addition of a further volume.)

4. Repeat steps *2* and *3* to obtain coincident results.

5. Record the laboratory temperature.

Calculations

Determine:

(*a*) the molarity of the diluted ammonium oxalate solution,

(*b*) the concentration of the undiluted ammonium oxalate in grams per litre,

(*c*) the solubility of ammonium oxalate at the laboratory temperature expressed in grams of solute per 100 cm^3 of solution.

Questions

1. What is the ionic equation for the reaction between ammonium oxalate and acidified potassium permanganate?

2. The figure obtained for the solubility of ammonium oxalate is not in accordance with the definition of solubility. What extra steps would be necessary to find the true solubility of ammonium oxalate?

2.2 Determination of the distribution coefficient of iodine between carbon tetrachloride and water

Theory

If a solute dissolves in two immiscible liquids, A and B, then

$$\frac{\text{concentration of the solute in solvent A}}{\text{concentration of the solute in solvent B}} = \begin{array}{l}\text{distribution coefficient}\\ \text{of the solute}\end{array}$$

as long as the solute is in the same molecular state in both solvents and neither solvent is saturated with the solute.

Requirements

60 cm^3 solution of iodine in carbon tetrachloride containing 7·0 g/l
50 cm^3 0·05M iodine solution
200 cm^3 c. 0·05M sodium thiosulphate solution
20 cm^3 carbon tetrachloride (TOXIC:CARE)
starch indicator
specified volumetric ware plus a 10 cm^3 pipette, a 100 cm^3 pipette and a 100 cm^3 graduated flask
250 cm^3 separating funnel
250 cm^3 stoppered bottle

Procedure

1. Fill two burettes, one with the solution of iodine in carbon tetrachloride and the other with carbon tetrachloride.†
2. Pipette 100 cm^3 of water into the separating funnel and proceed to make up one of the following mixtures (each member of the class should prepare a different mixture):

mixture	cm^3 of I_2 in CCl_4	cm^3 of CCl_4	cm^3 of water
A	50	0	100
B	48	2	100
·	·	·	·
J	32	18	100
K	30	20	100

Shake the mixture continuously for five minutes and allow the layers to separate.
3. Titrate 10 cm^3 portions of the 0·05M iodine solution with the c. 0·05M sodium thiosulphate solution. When the iodine solution

† Employ these two burettes communally for dispensing the two liquids.

reaches a pale straw colour, starch may be used as indicator. Repeat to obtain concordant results.†

4. Pipette 10 cm^3 of the 0·05M sodium thiosulphate solution into the 100 cm^3 graduated flask, make up to the mark with purified water and shake well to give a 0·005M solution.

5. Run the carbon tetrachloride layer from the separating funnel into a burette.

6. Burette 10 cm^3 of the carbon tetrachloride layer into the stoppered bottle and titrate, using 0·05M sodium thiosulphate solution; add 1·0 cm^3 at a time, shaking well after each addition, until the purple colour in the carbon tetrachloride is just removed.

7. Repeat the titration, but add the sodium thiosulphate solution dropwise over the crucial 1 cm^3 range, until the carbon tetrachloride layer is completely colourless.

8. Repeat step *7* twice more to obtain concordant results.

9. Discard the intermediate layer in the separating funnel and run the aqueous layer into a beaker.

10. Pipette two 20 cm^3 portions of the aqueous layer into a conical flask and carefully titrate with the 0·005M solution of sodium thiosulphate, using starch as indicator (the titre will be less than 10 cm^3). Repeat this titration.

11. Obtain results for the other mixtures from the various class experimenters.

12. Ascertain the laboratory temperature.

Observations and Calculations

1. Complete a table with the following headings:

mixture	g of iodine taken before distribution‡	CARBON TETRACHLORIDE LAYER			WATER LAYER			g of iodine present after distribution, i.e. A+2B‡	distribution ratio A/B
		cm^3 of 0·05M sodium thiosulphate solution for 10 cm^3 of layer	cm^3 of 0·05M sodium thiosulphate for total (i.e. 50 cm^3) layer	g of iodine present in 50 cm^3, i.e. A	cm^3 of 0·005M sodium thiosulphate for 40 cm^3 of layer	cm^3 of 0·005M sodium thiosulphate for 50 cm^3 of layer	g of iodine present in 50 cm^3 of layer, i.e. B		

† It is not essential to do this titration, but it provides a check on the total weight of iodine initially in solution.

‡ These two weights should agree.

2. From the class results plot a graph of A against B, and from the slope of the graph deduce the value of the distribution coefficient at the laboratory temperature.

Questions

1. If the values in columns 2 and 9 of the table differ, what explanation can be given?

2. In what way does the determined value of the distribution coefficient provide evidence as to the nature of the substances involved?

3. How may the different solubilities of iodine in water and carbon tetrachloride be mechanistically explained?

2.3 Determination of the solubility of sulphur dioxide in acetone *†

Requirements

4·5 g A.R. ammonium iron(II) sulphate
6 cm^3 saturated solution of sulphur dioxide in acetone
200 cm^3 c. 0·02M potassium dichromate solution
25 cm^3 syrupy orthophosphoric acid
150 cm^3 bench dilute sulphuric acid
barium diphenylamine sulphonate indicator solution
specified volumetric ware plus a 100 cm^3 graduated flask
ignition tube
test tube
plastic bowl

Procedure

1. Weigh out accurately about 3·9 g of the ammonium iron(II) sulphate in a 100 cm^3 beaker.

2. Dissolve the crystals in bench dilute sulphuric acid. Make up the acid solution and washings to 100 cm^3 in the graduated flask, using dilute sulphuric acid to give a standard solution (c. 0·1M). Shake the solution well.

3. Titrate 20 cm^3 portions of the ammonium iron(II) sulphate solution containing four drops of barium diphenylamine sulphonate indicator, with the potassium dichromate solution. When the colour turns dark green, add an ignition tube (c. 4 cm^3) of the

† Water may be used instead of acetone.

orthophosphoric acid and proceed to the end point. Obtain concordant titres.

4. Put about 200 cm^3 of water in a 250 cm^3 graduated flask, and add exactly 5·0 cm^3 of the saturated solution of sulphur dioxide in acetone.†

5. Immediately make up to the mark with purified water, shake well and stand in the bowl containing cold water.

6. Pipette 20 cm^3 of the dilute sulphurous acid into a conical flask, add half a test tube (c. 10 cm^3) of bench dilute sulphuric acid and four drops of barium diphenylamine sulphonate indicator, and titrate rapidly with the standardised potassium dichromate solution. Repeat the titration to give concordant results: this is difficult because the end-point colour tends to fade.

7. Record the laboratory temperature and barometric pressure.

Calculations

Determine:

(*a*) the molarity of the ammonium iron(II) sulphate solution,

(*b*) the molarity of the potassium dichromate solution,

(*c*) the molarity of the saturated sulphur dioxide solution (i.e. before dilution),

(*d*) the concentration in grams per litre of the saturated sulphur dioxide solution,

(*e*) the volume of sulphur dioxide at S.T.P. dissolved in 1·0 cm^3 of the solution,

(*f*) the volume of sulphur dioxide under the laboratory conditions dissolved in 1·0 cm^3 of the solution.

Questions

1. What is the ionic equation for the reaction between sulphur dioxide and potassium dichromate in acid solution?

2. The figure obtained for the volume of sulphur dioxide dissolved in 1·0 cm^3 of solution at S.T.P. is not in accordance with the definition for the Bunsen absorption coefficient. What extra steps would be necessary to find the true value for the absorption coefficient of sulphur dioxide?

3. What is the major source of inaccuracy in this experiment, and how might it be overcome?

† Dispense from a large burette which is used by the class.

4. Potassium permanganate is not as satisfactory an oxidant as potassium dichromate for the estimation of sulphur dioxide in an acetone/water solution. Why is this?

5. Given a solution of hydrogen peroxide, how could you use this to estimate the sulphur dioxide content of the saturated solution?

6. How might the solubility of carbon dioxide in alcohol be experimentally determined?

2.4 Determination of the solubility of ammonia in water

Requirements

400 cm³ 1·5M hydrochloric acid
110 cm³ c. 2·0M sodium hydroxide solution
20 cm³ 0·88 ammonia
phenolphthalein indicator solution
specified volumetric ware plus a 500 cm³ graduated flask
25 cm³ measuring cylinder
soft glass boiling tube fitted with a bung
1 lb jam jar

Procedure

1. Titrate 20 cm³ pipetted portions of the hydrochloric acid with the sodium hydroxide solution, using phenolphthalein as indicator. Repeat to obtain concordant results.

2. Fill up to the mark a 250 cm³ graduated flask with the hydrochloric acid, and transfer the acid with washings (do not employ more than 100 cm³ of purified water) to the jam jar.

3. Weigh the soft glass boiling tube and bung supported in a 100 cm³ beaker.

4. From the measuring cylinder, pour about 15 cm³ of the 0·88 ammonia into the soft glass boiling tube and pass in ammonia gas from a generator† for a few minutes to ensure maximum solubility at the laboratory temperature.

5. Place the rubber bung firmly into the boiling tube and reweigh as in step *3.*

6. Put the boiling tube into the jam jar and break it by sharply striking the bottom.

† A class generator can be set up by heating about 200 cm³ of 0·88 ammonia (containing a porous chip) in a 500 cm³ conical flask fitted with a suitable delivery tube.

7. Loosen the bung and stir the mixture well with the broken tube.
8. Transfer the liquid contents of the jam jar with washings to a 500 cm^3 graduated flask; make up to the mark with purified water.
9. From the burette, run 25 cm^3 of the standardised sodium hydroxide solution into a 250 cm^3 graduated flask, make up to the mark with purified water and shake well.
10. Titrate 20 cm^3 portions of the solution prepared in step *8*, using phenolphthalein as indicator, with the diluted standard sodium hydroxide solution, to obtain concordant results.
11. Record the barometric pressure and laboratory temperature.

Calculations

Determine:

(*a*) the molarity of the original sodium hydroxide solution,

(*b*) the number of moles of hydrochloric acid (*i*) taken initially in step *2*, (*ii*) in excess after addition of ammonia and (*iii*) neutralised by the ammonia,

(*c*) for the ammonia dissolved, (*i*) the number of moles, (*ii*) the volume at S.T.P. and (*iii*) the volume under the laboratory conditions,

(*d*) the Ostwald solubility coefficient for ammonia in water.

.5 Determination of the composition of the constant boiling mixture formed between hydrogen chloride and water

Theory

The temperature-composition curve for hydrochloric acid has the form:

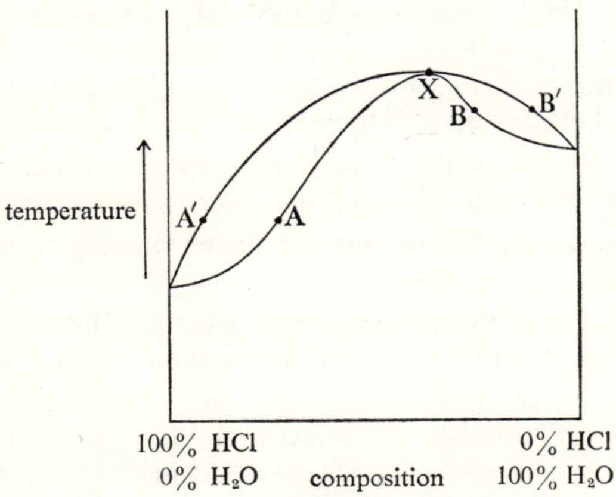

FIG. 2

When a solution of composition A is boiled, the resulting vapour of composition A′, in equilibrium with the solution at the same temperature, is richer in hydrogen chloride. The remaining solution has a greater proportion of water. On further boiling, with continuous removal of the vapour, the enrichment of the solution in water continues until the liquid and the vapour have the same composition and are in equilibrium, i.e. at point X. On boiling a solution of composition B, the situation is now reversed in that the vapour phase, B′, is richer in water than the liquid phase, B. Continued boiling of B with loss of vapour eventually moves the composition of the solution to X, where again the liquid and vapour are in equilibrium and have the same composition. In both cases, a constant boiling mixture of maximum boiling point is formed.

Requirements

100 cm^3 0·2M sodium hydroxide solution
60 cm^3 concentrated hydrochloric acid
methyl red indicator solution
specified volumetric ware plus a 500 cm^3 graduated flask
250 cm^3 distillation flask fitted with a one-holed rubber bung carrying a 0°C to 200°C thermometer
boiling tube
clean chip of porous pot

Procedure

1. Pour a boiling tube (c. 55 cm^3) of concentrated hydrochloric acid into the distillation flask, followed by the same volume of water.

2. Add the chip of clean porous pot and insert the bung so that the thermometer bulb is opposite the side arm of the distillation flask.

3. Boil the contents of the flask (preferably in a fume cupboard) until a steady temperature is reached; record this temperature.

4. Cool the constant boiling mixture under running water until it is about 25°C.

5. Pipette 20 cm^3 of the constant boiling mixture into the 500 cm^3 graduated flask. Make up to the mark with purified water and shake well.

6. Titrate 20 cm^3 portions of the sodium hydroxide solution with the diluted hydrochloric acid made in step *5* (methyl red as indicator) to obtain concordant results.

Calculations

Determine:

(*a*) the molarity of the diluted hydrochloric acid,

(*b*) the molarity of the constant boiling mixture,

(*c*) the concentration in grams per litre of the constant boiling mixture,

(*d*) the percentage composition of the constant boiling mixture (*i*) by weight and (*ii*) by mole fraction.

Questions

1. Why is the constant boiling mixture not a true compound? Refer to (*a*) the above experiment and (*b*) other sources.

2. Why do water and hydrogen chloride form a constant boiling mixture?

3. What is the exact interpretation of the curves (*a*) AXB and (*b*) A′XB′ in figure 2?

4. What is an alternative term for 'constant boiling mixture'?

6 Determination of the solubility of oxygen in water*

Theory

Excess manganese(II) hydroxide is added to a known volume of water saturated with oxygen; a portion of it (equivalent to the oxygen) is oxidised to manganese(III) hydroxide. This is dissolved in orthophosphoric acid, a non-oxidising acid, and allowed to oxidise iodide ions. The iodine, which is quantitatively liberated, is titrimetrically estimated with standard sodium thiosulphate solution.

Requirements

120 cm^3 water saturated with oxygen
30 cm^3 0·005M sodium thiosulphate solution
5 cm^3 c. 40 per cent manganese(II) chloride tetrahydrate solution
5 cm^3 solution containing c. 30 g sodium hydroxide and c. 10 g potassium iodide dissolved in 100 cm^3 of purified water
5 cm^3 syrupy orthophosphoric acid
starch indicator
specified volumetric ware plus a 50 cm^3 pipette
boiling tube with bung to fit†

† 50 cm^3 of water should be pipetted into the boiling tube and a bung selected which just makes contact with the surface of the water.

Procedure

1. Pipette 50 cm^3 of the oxygen solution into the boiling tube and then add, from microburettes, 0·5 cm^3 of manganese(II) chloride solution and 0·1 cm^3 of alkaline iodide solution.†

2. Ease the bung into the boiling tube so that contact is made with the surface of the liquid. Continuously and rapidly invert the boiling tube for not less than five minutes, and then leave to stand for at least a further five minutes.

3. Remove the bung and add, from a 50 cm^3 burette, 2 cm^3 of syrupy orthophosphoric acid. Replace the bung and repeatedly invert the boiling tube until a solution of iodine is obtained.

4. Pipette 20 cm^3 of this iodine solution into a conical flask and titrate with the sodium thiosulphate solution, dispensed from a microburette, using starch as indicator.

5. Repeat step *4*.

6. Repeat steps *1* to *5*.

7. Record the laboratory temperature and barometric pressure.

Questions

1. What are the ionic equations for the reactions involved in the estimation?

2. What is the solubility of oxygen in grams per cm^3 of water under the laboratory conditions?

3. (*a*) Why is it essential for the bung to make contact with the liquid contents of the boiling tube?

(*b*) What errors are there in the experiment, and how might they be overcome?

2.7 Determination of the solubility of aniline in aqueous solution*

Theory

Free aniline is liberated from aniline hydrochloride solution by the addition of sodium hydroxide:

$$C_6H_5NH_3Cl(aq) + NaOH(aq) \longrightarrow C_6H_5NH_2(aq/l) + NaCl(aq) + H_2O(l).$$

† Employ these two microburettes communally for dispensing the two solutions.

When sufficient aniline has been liberated to saturate the solution, the addition of more sodium hydroxide will cause the solution to become turbid owing to undissolved aniline.

If the volume of water present in the solution of aniline hydrochloride is increased, then more sodium hydroxide will be required in order to reach turbidity.

Requirements

110 cm^3 aniline hydrochloride solution
110 cm^3 1·0M sodium hydroxide solution
specified volumetric ware plus a 50 cm^3 burette and a 10 cm^3 pipette

Procedure

1. Put the 1·0M sodium hydroxide solution in one burette, and purified water in the other.

2. Pipette 10 cm^3 of the aniline hydrochloride solution into the conical flask and add sodium hydroxide solution until the solution just becomes turbid.

3. Record this result on a graph plotting the volume of sodium hydroxide solution added against the total volume of the solution at turbidity.

4. (*a*) Pipette 10 cm^3 of the aniline hydrochloride solution into a conical flask, add 3 cm^3 of water from the burette and titrate with the sodium hydroxide solution as before.
(*b*) Again, plot the result on the graph.

5. Repeat steps *4* (*a*) and (*b*), but this time add 4 cm^3 of water to the aniline hydrochloride solution.

6. Repeat step *5* a further six times, employing 5, 6, 7, 8, 9 and 10 cm^3 of water respectively, and thereby complete the graph.

Observations

Record the results in the following manner:

cm^3 of aniline hydrochloride	cm^3 of water added	cm^3 of sodium hydroxide added	total cm^3 at turbidity

Calculations

1. From the equation, determine the weight of aniline liberated by 1·0 cm^3 of sodium hydroxide solution.

2. From the slope of the graph, determine the volume of sodium hydroxide solution which will liberate sufficient aniline to saturate 10 cm^3 of aqueous solution.

Questions

1. What is the solubility of aniline at the laboratory temperature in grams per 100 cm^3 of water?

2. What is the nature of the turbidity observed during the experiment?

2.8 Establishment of the freezing point curve for a mixture of 4-toluidine and naphthalene*

Theory

When a cooling curve is plotted for a pure compound, there is one break in the curve, at the freezing point of that compound, but for a solution of two components, there are two breaks in the curve. The first one represents the temperature at which the solution begins to freeze, and the second one represents the freezing point of the eutectic mixture.

Requirements

50 g 4-toluidine
25 g naphthalene
250 cm^3 beaker
twelve test tubes and a two-holed rubber bung to fit all of them
−10°C to 110°C thermometer
stop-clock
wire stirrer

Proedure

1. Prepare test tubes with the following contents:

g of naphthalene	5·0	4·0	3·0	2·0	1·0	0
g of 4-toluidine	0	1·0	2·0	3·0	4·0	·0

2. Stand the first test tube in a beaker of hot water, remove when the contents are molten, and insert the bung carrying the thermometer and stirrer.

3. Stir continuously and record the temperature every minute.

4. Plot the cooling curve and continue it until the graph horizontally levels out (i.e. the eutectic temperature).

5. Repeat steps *2* to *4* for each test tube.

6. At this stage, carry out steps *1* and *2* of the next section, after which it will be apparent that there are insufficient points for a satisfactory freezing point curve.

7. Proceed to obtain cooling curves for further suitable mixtures.

Observations and Calculations

1. Tabulate the results in the following manner:

% composition		freezing point of mixture (first break in graph)	freezing point of eutectic (second break in graph)
4-toluidine	naphthalene		

2. For the samples, plot the percentage by weight of naphthalene against the freezing point.

3. From the graph, find:

   (*a*) the composition of the eutectic mixture,

   (*b*) the freezing point of the eutectic mixture.

4. Calculate the ratio of the number of moles in the eutectic mixture.

Question

The eutectic mixture is not a pure compound. What experimental evidence supports this statement?

3

THERMOCHEMISTRY

Introduction

Although chemical substances possess chemical energy, the absolute energy associated with any particular substance cannot be ascertained.

However, in chemical changes the difference in energy between reactants and products often manifests itself in either the liberation or the absorption of heat. Thermochemistry is a study of such heat changes, which when established experimentally help to throw light on the nature of the species involved in the reactions under investigation.

Throughout this chapter, $\triangle H$, the heat change of a chemical system, is negative when heat is evolved and positive when heat is absorbed.

3.1 Comparison of the heats of combustion of a series of normal alcohols

Theory

The heat of combustion of a substance is the quantity of heat liberated when one mole of the substance is completely burned in oxygen.

Requirements

30 cm^3 of each alcohol (methanol, ethanol, propan-1-ol and butan-1-ol) contained in separate 35 cm^3 glue bottles, each fitted with a metal screw cap carrying a lamp wick

250 cm^3 measuring cylinder

$-10°C$ to $110°C$ thermometer

two coffee tins (c. 7·5 cm diameter and 7·5 cm high), a one-holed plastic lid and three 3 cm long corks, assembled as shown in figure 3:

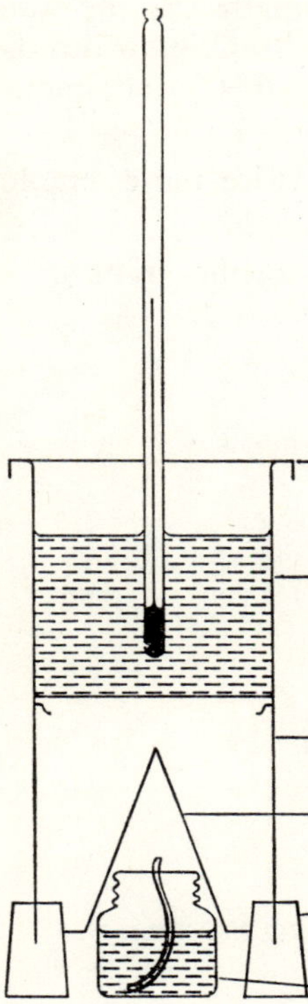

FIG. 3

Procedure

1. Using the 250 cm^3 measuring cylinder, put 200 cm^3 of water into the tin can.

2. Assemble all the apparatus with the exception of the thermometer and the glue bottle burner.

3. Take the methanol burner and ensure that the exposed wick is no longer than 0·5 cm.

4. Perform the following steps as rapidly as possible:

(*a*) put the thermometer through the plastic lid into the water, stir, and record the temperature of the water (i.e. θ°C),

(*b*) weigh the alcohol burner and put it in position,

(*c*) through a vent, light the wick.

5. With the thermometer, continually stir the water. When the temperature has risen very nearly 10°C, blow out the flame, record the exact temperature (i.e. $[\theta+\triangle\theta]$°C) and immediately reweigh the burner.

6. Repeat steps *1* to *5* at least twice more, employing the same alcohol.

7. Repeat steps *1* to *6* using each alcohol in turn.

Observations

Record the readings in the following way:

alcohol	temperature of water in °C		g of burner	
	initially θ	finally $(\theta+\triangle\theta)$	initially	finally

Calculations

Complete a table with the following headings:

alcohol	mole weight	$\triangle\theta$ in °C	g of alcohol burned	kcal gained by water	heat of combustion in kcal/mole

Conclusion

Record the average heat of combustion of each alcohol, and plot these quantities against the mole weights of the alcohols.

Questions

1. Look up and plot the accepted heats of combustion of the alcohols on the same axes. How do they compare with the experimental values?

2. From the graphs, what are (*a*) the experimental and (*b*) the accurate heat energies associated with the combustion of a CH_2 unit?

3. What are the drawbacks of this method, and how might some of them be overcome?

3.2 Investigation into the heats of neutralisation of various acids*

Theory

Heat of neutralisation may be defined as the quantity of heat evolved when one mole of water is produced from the neutralisation of an acid by a base.

Question

Is this definition acceptable? Reasons must be given.

Requirements

120 cm^3 1·0M hydrochloric acid
120 cm^3 1·0M nitric acid
120 cm^3 0·5M sulphuric acid
120 cm^3 1·0M acetic acid
240 cm^3 c. 1·1M sodium hydroxide
240 cm^3 c. 1·1M ammonium hydroxide
50 cm^3 pipette
−10°C to 110°C thermometer
two 250 cm^3 polythene bottles

Procedure

1. Pipette a 50 cm^3 portion of the hydrochloric acid into a polythene bottle and record the temperature when it is steady.

2. Similarly, pipette 50 cm^3 of the sodium hydroxide solution into the second polythene bottle and record the temperature when it is steady.

3. Mix the two solutions from steps *1* and *2* by pouring the alkali into the acid, and agitate by swirling. Record the maximum temperature reached.

4. Wash out both polythene bottles with water and allow them to drain for a minute or two.

5. Repeat steps *1* to *4* using the ammonium hydroxide in place of the sodium hydroxide.

6. Repeat steps *1* to *5*, using each of the other acids in turn.

Observations and Calculations

Taking the specific heats and densities of all the solutions to be unity, complete a table with the following headings:

acid employed	°C of acid before mixing	alkali employed	°C of alkali before mixing	maximum °C after mixing	calories evolved in the neutralisation of the acid with alkali	heat of neutralisation in kcal

Conclusion

From the heats of neutralisation, arrange the acids and alkalis in decreasing orders of strength.

Questions

1. Why is it necessary to have the solutions of the alkalis stronger than those of the acids?

2. The heats of neutralisation for some different pairs of acids and bases are of about the same value. Why is this?

3. For each acid, with the two alkalis, what is the difference in the two heats of neutralisation? What is the significance of these differences?

4. Why is it possible, in this experiment, to obtain satisfactory results employing polythene bottles in place of more conventional thermos flasks?

5. The heat of neutralisation of hydrofluoric acid is $-16{\cdot}3$ kcal. How can this result be adequately explained?

6. The heats of neutralisation for the chloro-acetic acids, $CH_2Cl.COOH$, $CHCl_2.COOH$ and $CCl_3.COOH$, increase with an increasing number of chlorine atoms in the molecule. Why is this?

3.3 Determination of the thermal capacity of a vacuum flask and fitments*

Theory

The quantity of heat lost by a mass of hot water when poured into a vacuum flask is equivalent to the quantity of heat gained by the flask and fitments:

$$\begin{pmatrix}\text{mass of}\\ \text{water in g}\end{pmatrix}\times\begin{pmatrix}\text{fall in temperature}\\ \text{of water in °C}\end{pmatrix}=C\times\begin{pmatrix}\text{rise in temperature of}\\ \text{flask and fitments in °C}\end{pmatrix}$$

where C is the thermal capacity of the vacuum flask and fitments expressed in calories per °C.

Requirements

100 cm^3 pipette
250 cm^3 beaker
-10°C to 110°C thermometer
vacuum flask fitted with another -10°C to 110°C thermometer (preferably immersion type) through a plastic lid and ring of rubber tubing, assembled as shown:

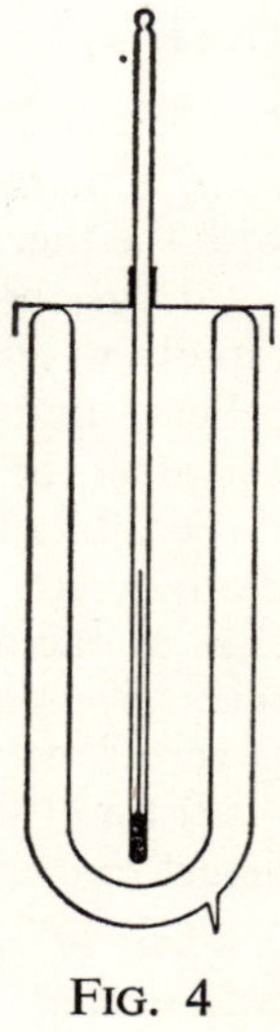

FIG. 4

Procedure

1. Expose both thermometers to the laboratory conditions for about two minutes. If the temperatures do not agree, an allowance must be made in subsequent calculations.

2. Set up the apparatus as illustrated and take the steady temperature of the interior of the vacuum flask.

3. Pipette 100 cm^3 of water into the beaker and, while stirring with the free thermometer, warm to about 30°C.

4. Stop heating, take the temperature of the warm water, rapidly pour it into the vacuum flask, and replace the plastic lid carrying the thermometer.

5. By careful use of the thermometer, stir the water in the vacuum flask and, by sensible withdrawal of the thermometer, note the maximum temperature attained.

Calculation

Determine the thermal capacity, in calories per °C, of the vacuum flask (with contents of volume 100 cm^3) and fitments.

3.4 Thermochemical investigation into the formation of potassium chloride

Requirements

20 g A.R. potassium hydroxide pellets
120 cm^3 c. 1·1M hydrochloric acid
60 cm^3 c. 2·2M hydrochloric acid
100 cm^3 pipette
50 cm^3 pipette
−10°C to 110°C thermometer
stop-clock
three pieces of glazed paper (c. 8 cm × 8 cm)
vacuum flask fitted with another −10°C to 110°C thermometer (preferably immersion type) through a plastic lid and ring of rubber tubing†

Procedure

1. Heat of solution of potassium hydroxide

(*a*) Pipette 100 cm^3 of purified water into the vacuum flask, stir with the thermometer, and when the temperature is steady record it.

(*b*) On one piece of glazed paper, with an open fold lengthways, rapidly weigh exactly 5·6 g (or slightly less) of potassium hydroxide pellets, immediately shoot the pellets into the vacuum flask and start the stop-clock.

(*c*) While continually stirring and by sensible withdrawal of the thermometer, record the temperature of the contents of the vacuum flask at one minute intervals for ten minutes; immediately plot each successive point on a graph of temperature against time, starting at zero time with the temperature recorded in step (*a*).

(*d*) Empty the vacuum flask, wash it out with purified water and invert to drain.

2. Heat of reaction between solid potassium hydroxide and hydrochloric acid

Repeat steps *1* (*a*) to (*d*), but employ a pipetted 100 cm^3 portion of c. 1·1M hydrochloric acid instead of purified water.

† As set up in experiment 3.3.

3. Heat of reaction between potassium hydroxide solution and hydrochloric acid

(*a*) Pipette 50 cm^3 of water into the vacuum flask.

(*b*) Repeat step *1* (*b*) (*i*) and (*ii*), and step *1* (*c*).

(*c*) After the tenth minute, employ the second thermometer to stir and take the temperature of the c. 2·2M hydrochloric acid; immediately pipette 50 cm^3 of it into the vacuum flask.

(*d*) Repeat step *1* (*c*).

Calculations

Take into account the thermal capacity of the vacuum flask and fitments,† and assume that the specific heat and density of all solutions are unity.

1. Heat of solution of potassium hydroxide

(*a*) Ascertain, by appropriate extrapolation, the maximum temperature difference.

(*b*) Calculate the heat of solution for 0·1 mole of potassium hydroxide in 100 cm^3 of water, and thus the heat change, $\triangle H_1$, for the following reaction:

$$\underset{0\cdot1\text{ mole}}{KOH(s)} + (aq)(100\text{ cm}^3) \rightarrow \underset{0\cdot1\text{ mole}}{KOH(aq)}\ (100\text{ cm}^3)$$

2. Heat of reaction between solid potassium hydroxide and hydrochloric acid

(*a*) Again ascertain the maximum temperature difference in the manner just outlined.

(*b*) Calculate the heat of reaction for 0·1 mole of solid potassium hydroxide with 100 cm^3 of c. 1·1M hydrochloric acid, and thus the heat change, $\triangle H_2$, for the following reaction:

$$\underset{0\cdot1\text{ mole}}{KOH(s)} + HCl(aq)\ (100\text{ cm}^3) \rightarrow \underset{0\cdot1\text{ mole}}{KCl(aq)}\ (100\text{ cm}^3) + H_2O(l)$$

3. Heat of reaction between potassium hydroxide solution and hydrochloric acid

(*a*) Assume that the temperature of the reactants, immediately before mixing, was the average of the steady temperature of the potassium hydroxide solution and the recorded temperature of the c. 2·2M hydrochloric acid, and ascertain the maximum temperature difference.

† As determined in experiment 3.3.

(*b*) Calculate the heat of reaction for 0·1 mole of potassium hydroxide and hydrochloric acid in 100 cm^3 of solution, and thus the heat change, $\triangle H_3$, for the following reaction:

$KOH(aq)$ (50 cm^3) + $HCl(aq)$ (50 cm^3) → $KCl(aq)$ (100 cm^3) + $H_2O(l)$
0·1 mole 0·1 mole

Conclusions

1. For 0·1 mole of acid, alkali and salt with 100 cm^3 as the final volume of salt solution, interpret the following Hess cycle:

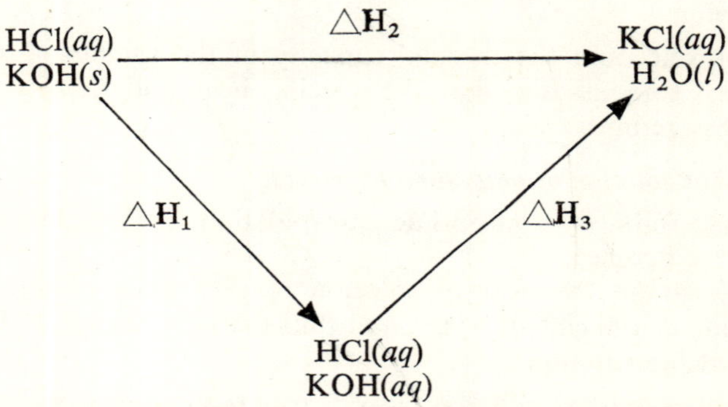

by substituting the experimentally determined heat quantities for $\triangle H_1$, $\triangle H_2$ and $\triangle H_3$.
2. Comment on the relationship between the experimentally determined heat quantities.

3.5 Verification of the Hess cycle by the formation of barium chloride

Requirements

2·2 g barium (TOXIC:CARE)
120 cm^3 0·2M hydrochloric acid
60 cm^3 0·4M hydrochloric acid
100 cm^3 pipette
50 cm^3 pipette
−5°C to 50°C (in 0·1°C) thermometer
stop-clock
wire cutters
vacuum flask fitted with another −5°C to 50°C (in 0·1°C) thermometer through a plastic lid and ring of rubber tubing†

† As set up in experiment 3.3.

Procedure

1. Check the thermometers as in experiment 3.3.

2. Heat of reaction between barium and hydrochloric acid

(*a*) With the wire cutters, cut a piece of clean dry barium about the size of a pea and weigh it (the mass should not exceed 1·0 g).

(*b*) Pipette 100 cm^3 of the 0·2M hydrochloric acid into the vacuum flask and, while continually stirring with the thermometer, plot a temperature against time graph for temperatures taken at half-minute intervals over a period of five minutes.

(*c*) Between the fifth and sixth minutes drop the weighed piece of barium into the vacuum flask. Continue stirring and plotting temperature against half-minute time intervals (i.e. a continuation of the previous graph) from the sixth minute onwards, for a further five minutes.

3. Heat of reaction between barium and water

Repeat steps *2* (*a*) to (*c*), employing a 50 cm^3 pipetted portion of purified water instead of the acid. Retain the contents of the vacuum flask for stage *4*.

4. Heat of reaction between barium hydroxide and hydrochloric acid

(*a*) Immediately after the eleventh minute, (*i*) use the second thermometer to stir and take the temperature of the 0·4M hydrochloric acid and (*ii*) pipette 50 cm^3 of it into the alkaline contents of the vacuum flask.

(*b*) Continue to obtain temperature-time data, in the manner of step *2* (*b*), and plot as a continuation of the graph from step *3*.

Calculations

Take into account the thermal capacity of the vacuum flask,† and assume that the specific heats and densities of the various reactants and products are unity. Relate all calculations to 1 mole of barium.

1. Heat of reaction between barium and hydrochloric acid

(*a*) By extrapolation of the graph, determine the maximum temperature difference.

(*b*) Calculate the heat of the reaction, $\triangle H_1$, for

$$Ba(s) + 2HCl(aq) \longrightarrow BaCl_2(aq) + H_2(g).$$

† As determined in experiment 3.3.

2. Heat of reaction between barium and water

In a manner similar to step *1* (i.e. by extrapolation and subsequent calculation), determine $\triangle H_2$ for the reaction

$$Ba(s)+2H_2O(l) \longrightarrow Ba(OH)_2(aq/s)+H_2(g).$$

3. Heat of reaction between barium hydroxide and hydrochloric acid

(*a*) Assume that the temperature of the reactants, immediately prior to mixing, was the mean of the temperature (by extrapolation) in the vacuum flask and the recorded temperature of the 0·4M hydrochloric acid, and ascertain the temperature difference for the reaction.

(*b*) Calculate the heat of reaction, $\triangle H_3$, for

$$Ba(OH)_2(aq/s)+2HCl(aq) \longrightarrow BaCl_2(aq)+2H_2O(l).$$

Conclusion

For 1 mole of barium, and with 100 cm^3 as the final volume of solution, complete the following Hess cycle:

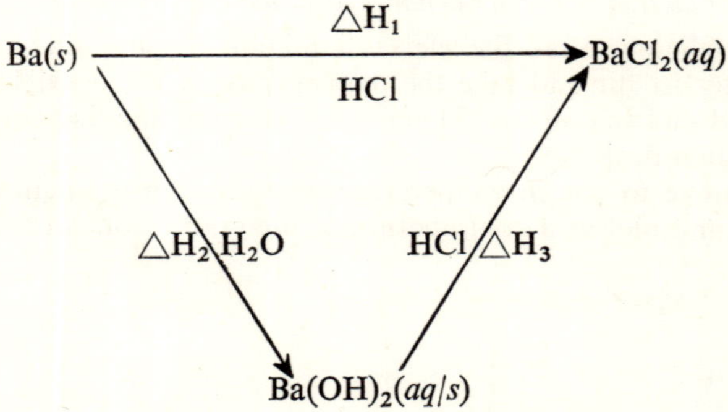

by substituting the experimentally determined values of $\triangle H_1$, $\triangle H_2$ and $\triangle H_3$. State whether Hess's Law is upheld.

Questions

1. The experiment involves a variation in the thermal capacity of the vacuum flask. Where does this occur and how might it be overcome?

2. What other sources of error are there in the experiment?

Determination of the heat of hydration of anhydrous copper(II) sulphate

Theory

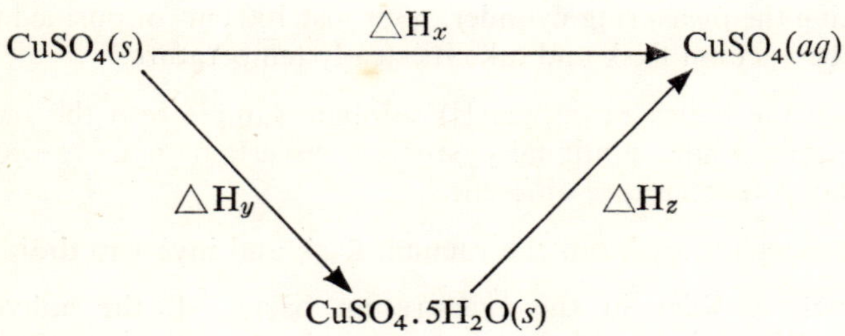

For the above Hess cycle, the heat quantities $\triangle H_x$ and $\triangle H_z$ may be determined experimentally and thus $\triangle H_y$, the heat of hydration of anhydrous copper(II) sulphate, deduced.

Requirements

60 g A.R. copper(II) sulphate crystals (fine)
vacuum flask fitted with a $-10°C$ to $110°C$ thermometer (preferably immersion type) through a plastic lid and ring of rubber tubing†
100 cm^3 measuring cylinder
100 cm^3 glass evaporating basin with a watch glass to fit in it as a lid
desiccator containing fused calcium chloride
small glass mortar and pestle
two pieces of glazed paper (c. 8 cm × 8 cm)

Procedure

1. With the mortar and pestle, very finely powder the copper(II) sulphate crystals.

2. Weigh the glass evaporating basin and then weigh exactly 25·0 g of the hydrated crystals into it.

3. Fit the watch glass in the evaporating basin so that the convex surface is above the crystals. Heat very gently until the top of the solid is white and the water which has condensed on the underside of the watch glass just disappears.

† As set up in experiment 3.3.

4. Transfer, in the evaporating basin, the prepared sample of anhydrous copper(II) sulphate to the desiccator.

5. On one piece of glazed paper, creased lengthways, weigh exactly 25·0 g of the copper(II) sulphate crystals.

6. Using the measuring cylinder, pour just 100 cm^3 of purified water into the vacuum flask and take its steady temperature.

7. Slide the hydrated copper(II) sulphate sample into the vacuum flask and, while continually stirring, ascertain and record the maximum temperature difference.

8. Thoroughly wash out the vacuum flask and invert to drain.

9. Finely powder in the evaporating basin, all the anhydrous copper(II) sulphate and then put it on the second piece of glazed paper.

10. Repeat steps *6* and *7*, but this time employ 109 cm^3 of purified water and all the sample of anhydrous copper(II) sulphate.

Observations

Maximum temperature difference for:

hydrated copper(II) sulphate = °C.

anhydrous copper(II) sulphate = °C.

Calculation

By application of the previously given diagrammatic cycle of Hess's Law, and making allowance for the thermal capacity of the vacuum flask and fitments†, calculate the heat of hydration of anhydrous copper(II) sulphate, expressing the heat quantity in kilocalories per mole.

Question

What is the significance of choosing (*a*) a 25·0 g sample of hydrated copper(II) sulphate and (*b*) 109 cm^3 of water?

† As determined in experiment 3.3.

7 Determination of the heat of decomposition of hydrogen peroxide in aqueous solution and the heat of formation of hydrogen peroxide

Requirements

1·0 g manganese(IV) oxide powder
220 cm^3 c. 2-volume hydrogen peroxide
100 cm^3 0·08M potassium permanganate solution
200 cm^3 bench dilute sulphuric acid
specified volumetric ware plus a 100 cm^3 pipette
vacuum flask fitted with a −10°C to 110°C thermometer (preferably immersion type) through a plastic lid and ring of rubber tubing†
boiling tube
stop-clock

Procedure

1. Estimation of the hydrogen peroxide solution

(*a*) Pipette a 20 cm^3 portion of it into a conical flask.

(*b*) Add two-thirds of a boiling tube (c. 40 cm^3) of bench dilute sulphuric acid.

(*c*) Titrate with the 0·08M potassium permanganate solution.

(*d*) Repeat steps (*a*) to (*c*) to obtain concordant results.

2. Investigation of the heat change involved in the decomposition of hydrogen peroxide, catalysed by manganese(IV) oxide, in aqueous solution

(*a*) Pipette 100 cm^3 of the hydrogen peroxide solution into the clean and dry vacuum flask, and assemble the apparatus.

(*b*) Start the clock and, while continually stirring with the thermometer, take and record (in a table) the temperature of the hydrogen peroxide solution, at minute intervals, for five minutes.

(*c*) Between the fifth and sixth minutes, shoot the whole sample of manganese(IV) oxide from the creased piece of glazed paper into the vacuum flask.

(*d*) Continue stirring and taking further temperatures (*i*) at the sixth minute, (*ii*) at minute intervals for a further ten minutes and (*iii*) finally at two five-minute intervals.

Calculations

1. Plot temperature against time for the decomposition and, by extrapolation, determine the maximum temperature rise.

† As in experiment 3.3.

2. Calculate the heat of decomposition for

$$H_2O_2(aq) \longrightarrow H_2O(l)+\tfrac{1}{2}O_2(g).$$

Neglect the thermal capacities of the manganese(IV) oxide and the evolved oxygen, but include that for the vacuum flask and fitments.† Assume that the density of the hydrogen peroxide solution and its specific heat are both unity.

3. Calculate the heat of formation of hydrogen peroxide from its gaseous elements by constructing a diagrammatic Hess cycle.‡ This should incorporate the heat of decomposition of hydrogen peroxide and the heat of formation of liquid water from its gaseous elements (i.e. $\triangle H = -68{\cdot}4$ kcal), but neglect the heat of solution of hydrogen peroxide.

† As determined in experiment 3.3.
‡ See experiment 3.4.

4

GAS DENSITY

Introduction

The density of a gas is expressed in two ways:

(*a*) Absolute density at S.T.P. $= \dfrac{\text{weight of gas}}{\text{volume of that weight of gas at S.T.P.}}$.

The usual units of absolute density are grams per cm^3 and, less frequently, grams per litre.

(*b*) Vapour density $= \dfrac{\text{weight of gas}}{\text{weight of equal volume of hydrogen at the same temperature and pressure}}$.

However, by Avogadro's hypothesis, equal volumes of gases at the same temperature and pressure contain the same number of molecules. Thus the previous expression becomes

$$\text{vapour density of a gas} = \frac{\text{weight of } x \text{ molecules of gas}}{\text{weight of } x \text{ molecules of hydrogen}}$$

$$= \frac{\text{weight of } N \text{ molecules of gas}}{\text{weight of } N \text{ molecules of hydrogen}}$$

where N is the number of carbon atoms in exactly 12 g of the carbon isotope ^{12}C (i.e. Avogadro's number, $6{\cdot}02246 \times 10^{23}$). Therefore,

$$\text{vapour density of a gas} = \frac{\text{weight of a mole of gas molecules}}{\text{weight of a mole of hydrogen molecules}}$$

$$= \frac{\text{weight of a mole of gas molecules}}{2 \times \text{weight of a mole of hydrogen atoms}}.$$

When both these weights are expressed on the carbon isotope, ^{12}C, scale (i.e. as in the I.U.P.A.C. *Table of Relative Atomic Masses 1961*), then the

$$\text{vapour density of a gas} = \frac{\text{mole weight of gas molecules}}{2 \times 1{\cdot}008}$$

i.e. for the gas molecules,

$$\text{mole weight} = 2{\cdot}016 \times \text{vapour density.}$$

Furthermore, the mole weight of a gas (or indeed any compound) is the weight in grams of Avogadro's number of molecules of the gas (or compound), and the volume occupied by Avogadro's number of molecules of an ideal gas at S.T.P. is 22·4 litres (i.e. the molar volume).

Questions

1. The weight of a nitrogen molecule is $4{\cdot}651 \times 10^{-23}$ g. What is the absolute density of nitrogen in grams per cm^3 at S.T.P.?

2. The weight of an oxygen atom is $2{\cdot}656 \times 10^{-23}$ g. What is the molar volume of oxygen at 25°C and one atmosphere pressure?

3. If the number of molecules in one mole of nitrogen were found to be 10^{20}, what would be the mole weight of nitrogen atoms?

4. The molar volume of all gases is not always exactly 22·4 litres at S.T.P. Why is this?

5. The preceding introductory theory must be amended for a noble gas. Why is this?

4.1 Mole weight of a gas by an adaptation of Regnault's method*

Requirements

supply of a suitable gas

500 cm^3 measuring cylinder

c. 500 cm^3 plastic bottle fitted with (*a*) a copper wire band and loop to facilitate weighing, and (*b*) a rubber bung through which passes a rigid plastic delivery tube attached to a rubber tube to permit the application of a Hofmann clip as illustrated in figure 5.

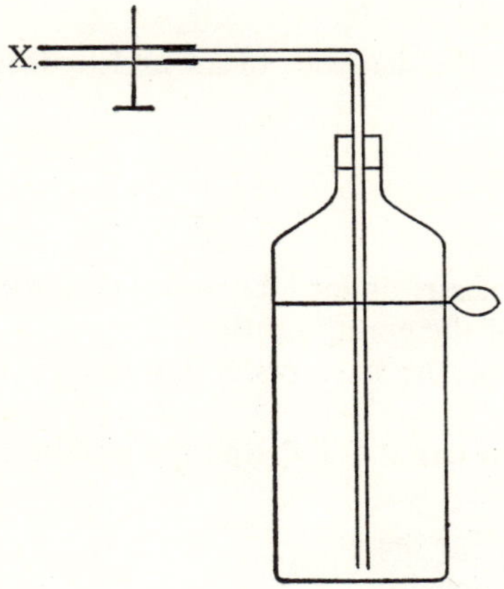

FIG. 5

Procedure

1. Weigh the dry plastic bottle (full of air) and dry attachments (i.e. bung, plastic and rubber delivery tubes, Hofmann clip and copper wire).

2. With the bung slightly lifted, attach a gas supply at X. Slowly pass the gas into the plastic bottle for at least five minutes, to ensure that all the air has been displaced.

3. Push the bung into the neck, close the Hofmann clip, and weigh the plastic bottle (full of gas) and attachments.

4. Repeat steps *2* and *3* to obtain two weighings which agree.

5. Record the laboratory temperature and the barometric pressure.

6. Fill the plastic bottle and the delivery tube up to the clip with water. Empty this water into the 500 cm^3 measuring cylinder and ascertain its volume.

Observations

Weight of dry bottle (full of air) and attachments	=	g
Weight of dry bottle (full of gas) and attachments (*a*)	=	g
(*b*)	=	g
etc.		
Barometric pressure	=	cm Hg
Laboratory temperature	=	°C
Volume of water to fill the bottle and delivery tube	=	cm^3

Data

From Tables, find the densities of air and hydrogen at S.T.P.

Calculations

1. Determine:

(*a*) the weight of air, under laboratory conditions, in the bottle,
(*b*) the weight of the empty bottle,
(*c*) the weight of the gas, under laboratory conditions, in the bottle,
(*d*) the volume of gas at S.T.P. and the weight of an equal volume of hydrogen.

2. Then calculate, for the gas:

(*a*) the absolute density,
(*b*) the vapour density,
(*c*) the mole weight.

4.2 Mole weight of a volatile liquid

A *By an adaptation of Dumas' method**

Requirements

10 cm^3 of a volatile liquid
unadjusted specific gravity bottle of c. 100 cm^3 capacity
50 cm^3 burette
250 cm^3 beaker
$-10°C$ to $110°C$ thermometer
ignition tube
boiling tube

Procedure

1. Dry the specific gravity bottle and stopper, and weigh them to 0·001 g.

2. Add half an ignition tube (c. 2 cm^3) of the pure volatile liquid to the bottle and insert the stopper.

3. Heat the water in the beaker and, when boiling, immerse the bottle in it as shown in the diagram. Keep the neck of the bottle above, and the mouth of the test tube just below, the level of the water.

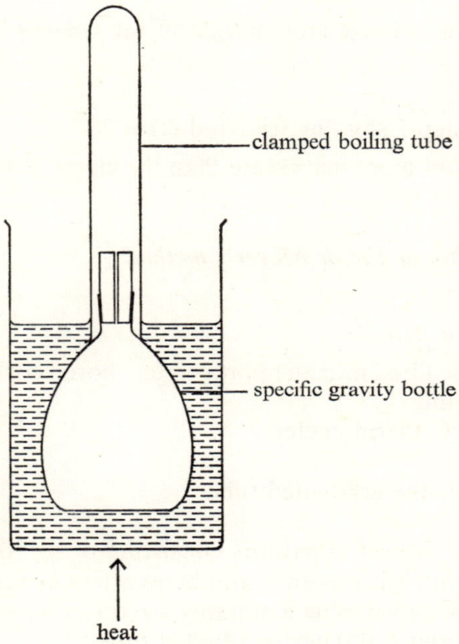

FIG. 6

4. About two minutes after the vapour has ceased to issue from the stopper, take the temperature of the water bath, remove the bottle from the water and carefully dry it.

5. When the bottle is *cold*, reweigh it.

6. Repeat steps *2* to *5* to obtain concordant results.

7. Use the burette to measure (by trial and error) the volume of the stoppered bottle to within 0·1 cm^3.

8. Record the barometric pressure.

Observations and Calculations

1. Volume of the specific gravity bottle = cm^3
Weight of bottle full of air = g
Barometric pressure = cm Hg

2. Complete a table with the following headings:

g of bottle full of air + condensed vapour	g of condensed vapour	°C of hot vapour	cm^3 of vapour at S.T.P.	g of vapour which would occupy 22·4 l at S.T.P.

3. Calculate the average mole weight of the volatile liquid.

Questions

1. Why does step *7* say, 'by trial and error'?

2. Is this method more inaccurate than the classical method? If so, why?

*B By an adaptation of Victor Meyer's method**

Requirements

4 cm^3 volatile liquid
Hofmann bottle fitted into an appropriately bored cork (as a support during weighing)
$-10°C$ to $110°C$ thermometer
teat pipette
gas jar as tall as the graduated tube
plastic bucket
modified Victor Meyer apparatus consisting of a 250 cm^3 Buchner flask fitted with various tube and bung attachments, and a Mohr spring clip as shown, plus a suitably sized can as a water bath, a plastic bowl and a 100 cm^3 graduated tube:

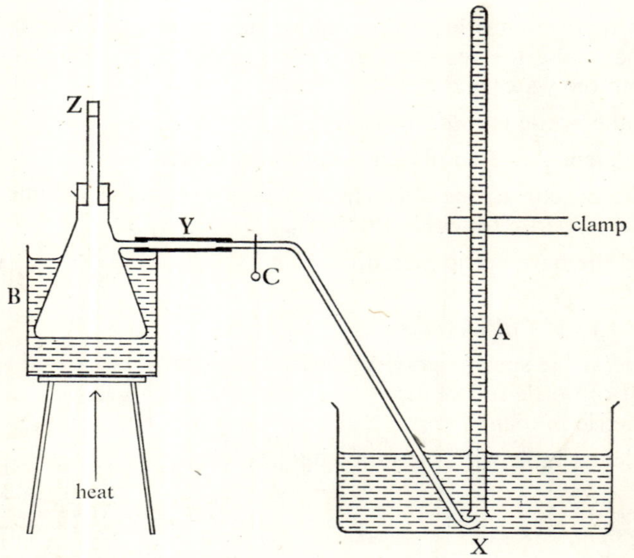

FIG. 7

Procedure

1. Ensure that the interior of the Buchner flask and attachments are dry, and then assemble the apparatus without the graduated tube A in position.

2. Heat the water in the bath B until bubbles cease to come from the delivery tube at X, and then keep the water in B gently boiling.

3. In the meantime, (*a*) weigh the empty, dry stoppered Hofmann bottle in the cork support; (*b*) with the aid of the teat pipette, fill the Hofmann bottle with the volatile liquid; and (*c*) weigh the stoppered Hofmann bottle, full of liquid, in the cork support.

4. Fill A and the gas jar full of water. Clamp A in position and stand the gas jar in the empty plastic bucket.

5. Quickly (*a*) simultaneously apply the Mohr spring clip C at Y and remove the bung at Z, (*b*) drop the Hofmann bottle into the Buchner flask and (*c*) replace the bung at Z and then immediately open the clip at Y.

6. When no more displaced air collects in A, close, underwater, the mouth of it with the thumb (or bung) and transfer A, without inverting it, into the gas jar. Hold A vertically and submerge it in the water. After about three minutes, raise A until the water levels in the two vessels coincide, and proceed to measure the volume of air in A.

7. Take and record the temperature of the water in the gas jar.

8. Remove the Hofmann bottle from the Buchner flask and make sure the bottle is dry.

9. Repeat steps *1* to *7* at least twice more.

10. Record the barometric pressure.

Observations, Data and Calculations

1. Temperature of the water in the gas jar = °C
S.V.P. of water at this temperature (from Tables) = cm Hg
Barometric pressure = cm Hg

2. Complete a table headed thus:

g of stoppered bottle and cork support	g of stoppered bottle, liquid and cork support	g of liquid	cm^3 of air collected	cm^3 of air and also vapour at S.T.P.	g of vapour occupying 22·4 l at S.T.P.

3. Calculate the average mole weight of the volatile liquid.

Questions

1. Why is it necessary to carry out instruction *2* of the procedure?

2. Why is the graduated tube containing displaced air held under water for three minutes?

3. What are the sources of error in this experiment?

4.3 Investigation of the relationship between the rate of effusion and the vapour density of a gas*

Theory

Graham's Law of Gaseous Diffusion states that the rate of diffusion of a gas is inversely proportional to the square root of its density. The Law also applies to effusion which takes place when a gas, under pressure, passes through a small orifice. Hence, the times taken (t) for equal volumes of gases to diffuse under the same conditions are directly proportional to the square roots of the densities (d);

i.e. $$t \propto \sqrt{d} \text{ or } t^2 \propto d$$

or in terms of vapour density (*v.d.*):

$$t \propto \sqrt{v.d.} \text{ or } t^2 \propto v.d.$$

Requirements

supplies of various dry gases (oxygen, carbon dioxide, sulphur dioxide, acetylene, butane, town gas, air)
100 cm^3 syringe fitted with a Luer needle holder (the hypodermic needle should be pulled out of the metal holder)
thin disc of cork (c. 1·5 cm diameter)
stop-clock

Procedure

1. Assemble the syringe with the disc of cork on top of the plunger (the small disc of cork facilitates the reading of the volume effused).

2. Mount the syringe vertically by clipping or clamping the neck of the plunger.

3. Connect a dry gas supply to the syringe nozzle and fill the syringe with the gas.

4. Disconnect the gas supply and allow the gas to effuse.

5. Repeat steps *3* and *4* at least twice more, to sweep out the syringe with the gas.

6. Fill the syringe with gas.

7. Disconnect the gas supply and simultaneously cover the nozzle with the tip of the first finger.

8. Slip the needle holder onto the nozzle.

9. Time the effusion of 90 cm^3 of gas from the 100 cm^3 graduation to the 10 cm^3 graduation.

10. Repeat steps *6* to *9* to obtain concordant times.

11. Repeat steps *3* to *9* for as many dry gases as available.

Observations and Calculations

1. Complete a table with the following headings:

gas	sec. for effusion of 90 cm^3 of gas		mole weight of the gas	vapour density† of the gas
	t	t^2		

2. Plot a graph of t^2 against vapour density for the data obtained.

† For an impure gas, the density may be determined by the adaptation of Regnault's method (see experiment 4.1).

5

MOLE WEIGHT OF SOLUTES

Introduction

Subject to certain limitations, the colligative properties of dilute solutions are directly proportional to the number of solute molecules in a given weight of solvent.

Hence, for two solutes, A and B, in the same weight of solvent which produces identical colligative effects, the weight of a mole of A can be found if the weight of a mole of B is known, as

$$\frac{\text{weight of A}}{\text{mole weight of A}} = \frac{\text{weight of B}}{\text{mole weight of B}}.$$

Question

What are the limitations referred to above?

5.1 Determination of the mole weight of acetamide by the depression of the freezing point of acetic acid*

Theory

For a dilute solution, the relationship between the depression of the freezing point and the mole concentration is given by

depression of freezing point = constant × mole concentration.

For a solution in acetic acid, with the mole concentration being expressed in mole/1000 g of acetic acid, the freezing point constant (cryoscopic constant) for acetic acid is 3·90°C, so that

$$\triangle\theta = 3{\cdot}90 \times \frac{c}{M}$$

where, for any particular solution in acetic acid,
$\triangle\theta$ is the depression of the freezing point in °C,
c is the concentration of solute in g/1000 g of acetic acid,
and M g is the mole weight of the solute.

Requirements

2·5 g acetamide
30 cm^3 glacial acetic acid
ice
50 cm^3 burette
$-10°C$ to 110°C thermometer
stop-clock
hand lens
four pieces of glazed paper (c. 8 cm × 8 cm)
eight pieces of white blotting paper (c. 8 cm × 8 cm) or filter paper
350 cm^3 plastic carton and lid, boiling tube and a two-holed cork, a $-5°C$ to 50°C (in 0·1°C) thermometer and a wire stirrer, assembled as shown:

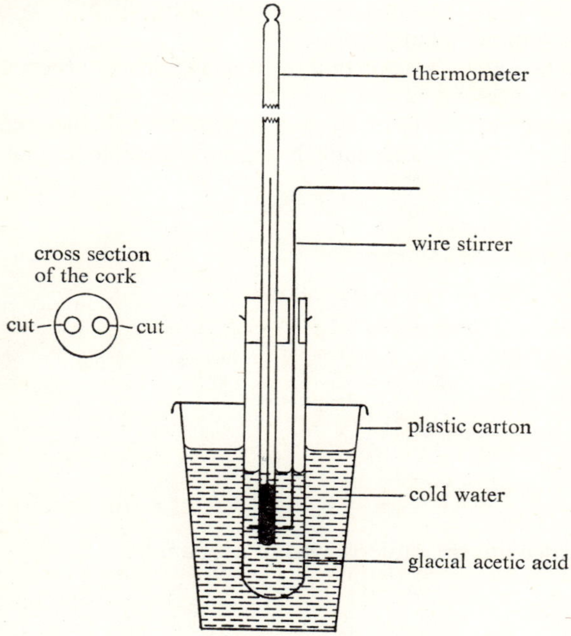

FIG. 8

Procedure

1. Fill the plastic carton with water, about two-thirds full, and adjust the temperature of the water to about 10°C (adding pieces of ice, if necessary).

2. From the burette, run exactly 25 cm^3 of the acetic acid into the boiling tube, and assemble the whole apparatus.

3. While continually stirring, (*a*) with the lens read the temperature to 0·02°C every minute and (*b*) immediately plot each point on a cooling curve of temperature against time.
4. Continue steps *3* (*a*) and (*b*) until the freezing point of acetic acid is found (supercooling usually takes place, and hence the temperature will rise to the freezing point of acetic acid and then slowly fall).
5. Raise the boiling tube out of the water to allow the acetic acid to melt, and readjust the temperature of the water to about 7°C.
6. Dry some acetamide by pressing between two pieces of blotting paper. On a piece of glazed paper, creased lengthways, weigh accurately about 0·5 g of the dry acetamide.
7. Remove the fittings from the boiling tube, slide the acetamide into the acetic acid, and ensure complete solution by use of the stirrer. Reassemble the apparatus.
8. Repeat steps *3* and *4* to ascertain the freezing point of the solution of acetamide in acetic acid.
9. By successive repetition of steps *5* to *8*, obtain further readings until a total of 2 g of acetamide has been added to the original 25 cm^3 of acetic acid.

Observations and Calculations

1. Draw up a table of results as follows:

cm^3 of glacial acetic acid	g of acetic acid, i.e. $cm^3 \times$ density (1·05 g/cm^3)	g of acetamide	conc. of acetamide in g/1000 g of acetic acid	freezing point in °C	$\Delta\theta$ in °C	M, i.e. $3{\cdot}90 \times \frac{c}{\Delta\theta}$

2. Since $c = \frac{M}{3{\cdot}90} \times \triangle\theta$, plot c against $\triangle\theta$, and from the slope of the graph, calculate the mole weight of acetamide.

Questions

1. If the acetamide smells of mice, (*a*) what impurity is present, (*b*) what effect would it have on the freezing point depression and (*c*) what effect would it have on the experimentally determined mole weight of acetamide?

2. Water and benzene are recommended solvents for cryoscopy. What are the relative advantages and disadvantages of these two solvents compared with glacial acetic acid?

2 Comparison of the effects of urea and potassium chloride on the boiling point of water

Theory

For a dilute solution, the relationship between the elevation of the boiling point and the mole concentration is given by

elevation of boiling point = constant × mole concentration.

For an aqueous solution, with the mole concentration being expressed in mole/1000 g of water, the boiling point constant (ebullioscopic constant) for water is 0·512°C, so that

$$\triangle\theta = 0{\cdot}512 \times \frac{c}{M}$$

where, for any particular aqueous solution,
$\triangle\theta$ is the elevation of the boiling point in °C,
c is the concentration of solute in g/1000 g of water,
and M g is the mole weight of the solute.

Requirements

6·0 g A.R. urea
6·0 g A.R. potassium chloride
hand lens
twelve pieces of glazed paper (c. 8 cm × 8 cm)
Landsberger apparatus and a −5°C to 105°C (in 0·1°C) thermometer, assembled as shown, together with a suitable steam generator:

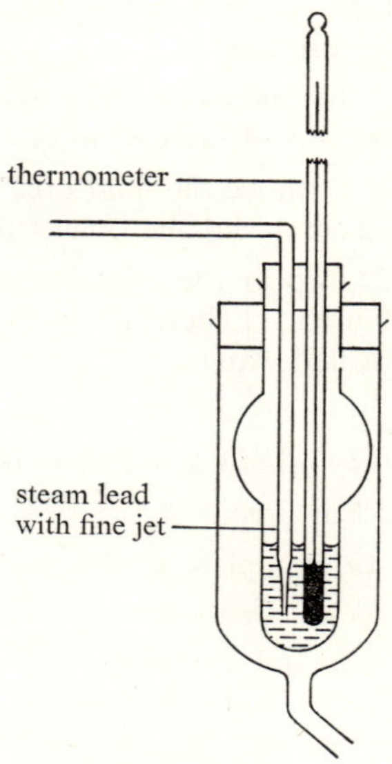

FIG. 9

Procedure

1. Put about 10 cm^3 of purified water in the inner graduated tube of the Landsberger apparatus.

2. Connect the steam lead to the steam generator and pass steam until the temperature reaches a steady value at the boiling point of purified water.

3. Using the hand lens, read the temperature. Record this temperature and disconnect the steam supply.

4. On a piece of glazed paper, creased lengthways, weigh accurately about 1 g of urea.

5. Remove the thermometer and steam lead from the inner tube and adjust the volume of water to about 10 cm^3.

6. Shoot the portion of urea down the middle of the inner tube.

7. Replace the thermometer and steam lead, and connect up the steam generator. Raise the temperature of the solution to its boiling point.

8. Using the hand lens, read the temperature until a maximum is reached. Record this temperature and immediately disconnect the steam supply.

9. Raise the thermometer and steam lead so that they are just out of the solution, allow them to drain and then record the volume of the solution.

10. By successive repetition of steps *4* to *9*, add four further 1 g portions of urea without discarding any solution in step *5*.

11. Remove the final urea solution from the Landsberger apparatus and wash out the inner tube with purified water.

12. Repeat the experiment employing potassium chloride in place of urea. (There is no need to re-determine the boiling point of purified water.)

Observations and Calculations

1. For urea,

(*a*) complete a table with the following headings:

g of water	g of urea	conc. of urea in g/1000 g of water	boiling point in °C	$\Delta\theta$ in °C	M, i.e. $0{\cdot}512 \times \frac{c}{\Delta\theta}$

(*b*) since $c = \dfrac{M}{0{\cdot}512} \times \triangle\theta$, plot *c* against $\triangle\theta$, and from the slope of the curve determine the mole weight of urea.

2. For potassium chloride,

(*a*) complete a table similar to the above,
(*b*) plot *c* against $\triangle\theta$.

Questions

1. In step *8*, why is a steady temperature not attained?

2. (*a*) What is the observed mole weight of potassium chloride?
(*b*) How can this value be explained?

3. (*a*) How would the apparatus below be used to determine the mole weight of naphthalene?

(*b*) What are the advantages, if any, of this apparatus compared with the more conventional Landsberger apparatus?

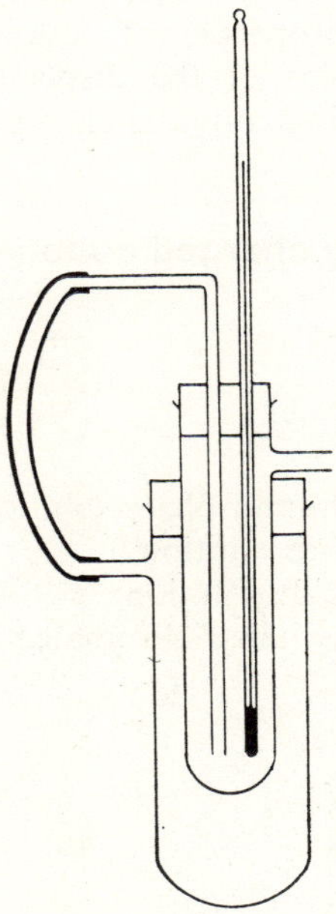

FIG. 10

6

COLLOIDS

Introduction

A colloid is a two-phase system in which small particles of a disperse phase are dispersed throughout a dispersion medium. The particles have a dimension range of about 10^{-7} cm to 10^{-5} cm.

Particles within a one-phase system are wholly surrounded by the same particle environment and are all subjected to the same forces of attraction. This is not true for all the particles within a two-phase colloidal system; the particles at the boundaries of the phases exert forces of attraction on each other which are unbalanced.

The special interfacial properties of colloids coupled with the relatively large surface areas of the dispersed particles play an important part in determining colloidal characteristics.

6.1 Preparation of positively charged colloids

A Iron(III) hydroxide

Requirements

0·6 g iron(III) chloride
20 cm^3 0·2M trisodium orthophosphate solution
20 cm^3 0·2M sodium sulphate solution
100 cm^3 0·2M sodium chloride solution
250 cm^3 filtered bench potassium dichromate solution
50 cm^3 burette
two 250 cm^3 beakers
250 cm^3 conical flask
glass stirring rod
ignition tube
test tube
film-strip projector fitted in the slide carrier with a piece of cardboard (5 cm$\times$5 cm) which is centre-punched (by cork borer) with an approximately 0·3 cm diameter hole

Procedure

1. Dissolve one-third of an ignition tube (c. 0·5 g) of iron(III) chloride in half a test tube (c. 10 cm^3) of purified water.

2. Put about 250 cm^3 of purified water in a beaker and boil. Hydrolyse the iron(III) chloride solution by pouring it slowly, with constant stirring, into the boiling water.

3. Away from sunlight and preferably in a dark room, pass the restricted beam of light from the film-strip projector through (*a*) the contents of the beaker and (*b*) a beaker full of the potassium dichromate solution. Compare the two effects.

4. (*a*) Put a test tube (c. 20 cm^3) of the iron(III) hydroxide colloid into the conical flask.

(*b*) Put the trisodium orthophosphate solution in the burette and determine the volume of it required to cause coagulation (indicated by the first trace of cloudiness).

5. Repeat step *4* using the solutions of (*a*) sodium sulphate and (*b*) sodium chloride instead of trisodium orthophosphate solution.

6. Retain the residual colloid for experiment 6.3.

Questions

1. What is the equation for the hydrolysis?

2. What explanations can be given for the observations in steps *3*, *4* and *5*?

B *Silver(I) iodide*

Requirements

55 cm^3 0·01M silver(I) nitrate solution
15 cm^3 0·025M potassium iodide solution
20 cm^3 0·2M trisodium orthophosphate solution
20 cm^3 0·2M sodium sulphate solution
100 cm^3 0·2M sodium chloride solution
50 cm^3 burette
100 cm^3 beaker
250 cm^3 conical flask
test tube
boiling tube
glass stirring rod

Procedure

1. Pour a boiling tube (c. 55 cm^3) of the silver(I) nitrate solution into the beaker.

2. Slowly add, with vigorous stirring, three-quarters of a test tube (c. 15 cm^3) of the potassium iodide solution.

3. Repeat steps *4* to *6* of experiment 6.1A.

Questions

1. What explanation can be given for the charge on the colloid?

2. How could a negatively charged colloid of silver(I) iodide be prepared?

6.2 Preparation of negatively charged colloids

A Arsenic(III) sulphide

Requirements

0·6 g arsenic(III) oxide (TOXIC:CARE)
20 cm^3 0·2M iron(III) chloride solution
20 cm^3 0·2M magnesium chloride solution
100 cm^3 0·2M sodium chloride solution
250 cm^3 filtered bench potassium chromate solution
50 cm^3 burette
two 250 cm^3 beakers
250 cm^3 conical flask
ignition tube
test tube
boiling tube
glass stirring rod
hydrogen sulphide generator fitted with a wash bottle containing water
film-strip projector and fittings†

Procedure

1. Put an eighth of an ignition tube (c. 0·5 g) of arsenic(III) oxide into about 250 cm^3 of purified water in a beaker. Boil and then allow to cool.

2. Half fill a boiling tube with this solution and keep for later use.

† See experiment 6.1A.

3. Pass hydrogen sulphide through the bulk of the solution until it smells of this gas.

4. Add the solution retained in step *2* to the contents of the beaker and stir.

5. In darkened surroundings, pass the restricted beam of light from the film-strip projector through (*a*) the contents of the beaker and (*b*) a beaker full of the potassium chromate solution. Compare the two effects.

6. (*a*) Put a test tube (c. 20 cm^3) of the arsenic(III) sulphide colloid into the conical flask.

(*b*) Put the iron(III) chloride solution in the burette and determine the volume of it required to cause coagulation (indicated by the first trace of cloudiness).

7. Repeat step *6* using the solutions of (*a*) magnesium chloride and (*b*) sodium chloride.

8. Retain the residual colloidal arsenic(III) sulphide for experiment 6.3.

Questions

1. What is the equation for the preparation?

2. What is the reason for carrying out step *4*?

3. What explanation can be given to the observations in steps *5* to *7*?

B *Antimony(III) sulphide*

Requirements

2·5 g antimony potassium tartrate (TOXIC:CARE)
3 cm^3 yellow ammonium sulphide
three 100 cm^3 beakers
250 cm^3 beaker
ignition tube
test tube
two boiling tubes

Procedure

1. Dissolve half an ignition tube (c. 2·0 g) of antimony potassium tartrate in a 100 cm^3 beaker full of purified water.

2. Pour half an ignition tube (c. 2 cm^3) of yellow ammonium sulphide into a 100 cm^3 beaker full of purified water.

3. Mix about half a boiling tube (c. 25 cm^3) of each of the solutions prepared in steps *1* and *2* in the third 100 cm^3 beaker. Retain the precipitate as a comparative standard.

4. (*a*) Add half a test tube (c. 10 cm^3) of the antimony potassium tartrate solution from step *1* to a boiling tube and fill up with purified water.

(*b*) Repeat (*a*) to obtain the same volume of a similarly diluted solution of ammonium sulphide.

(*c*) In the 250 cm^3 beaker, mix these diluted solutions, observe and compare with the standard from step *3*. If the product is turbid, then repeat, using more dilute solutions in (*a*) and (*b*).

5. Subject half a boiling tube of the non-turbid product to prolonged boiling and observe.

6. Retain the colloidal product for experiment 6.3.

Questions

1. What is the formula and common name of antimony potassium tartrate?

2. What is the equation for the preparation of the colloidal antimony(III) sulphide?

3. What explanation may be given for the observation in step *5*?

4. What is the chemical composition of yellow ammonium sulphide?

6.3 Mutual coagulation of colloids

Requirements

100 cm^3 iron(III) hydroxide sol†
100 cm^3 arsenic(III) sulphide sol‡ (TOXIC: CARE)
two 50 cm^3 burettes
250 cm^3 conical flask

Procedure

1. Put a sol into each burette.

2. (*a*) Run 20 cm^3 of arsenic(III) sulphide sol into the conical flask.

(*b*) Add iron(III) hydroxide sol to produce slight cloudiness and record the volume used.

3. (*a*) Run 20 cm^3 of iron(III) hydroxide sol into the conical flask.

† Alternatively, silver(I) iodide sol from experiment 6.1B.

‡ Alternatively, antimony(III) sulphide sol from experiment 6.2B.

(*b*) Add just sufficient arsenic(III) sulphide sol to cause cloudiness and record this volume.

Questions

1. In each case, what is the cause and explanation of the cloudiness?

2. What significance is there in the volumes employed?

6.4 Various preparations of colloidal sulphur

A Physical methods

1 ***By grinding***

Requirements

2·5 g flowers of sulphur
2·5 g urea
mortar and pestle
filter funnel
two ignition tubes
test tube
boiling tube

Procedure

1. Put half an ignition tube (2·0 g) each of flowers of sulphur and urea into the mortar and grind for at least ten minutes with the pestle, until extremely fine.

2. Add a test tube (c. 20 cm^3) of water to the powder and mix well.

3. Filter the suspension into the boiling tube and observe.

Question

Why is urea employed in this experiment?

2 ***By variation of solubility with solvent***

Requirements

2·5 g crushed roll sulphur
25 cm^3 industrial alcohol
500 cm^3 beaker
filter funnel
ignition tube
test tube
two boiling tubes

Procedure

1. Two-thirds fill the beaker with water, bring to the boil and then turn out the Bunsen burner.

2. Put a test tube (c. 20 cm^3) of the alcohol and half an ignition tube (2·0 g) of crushed roll sulphur into a boiling tube. Stand the boiling tube in the beaker of hot water for about two minutes.

3. Filter off the undissolved sulphur, using the second boiling tube to collect the filtrate.

4. Nearly fill the beaker with cold water and add the alcoholic solution of sulphur.

Question

Why is roll sulphur soluble in alcohol yet insoluble in water?

B Chemical methods

1 By reaction between hydrogen sulphide and sulphur dioxide in aqueous solution

Requirements

hydrogen sulphide generator
sulphur dioxide canister
filter funnel
two boiling tubes

Procedure

1. Pass a slow stream of hydrogen sulphide into half a boiling tube (c. 25 cm^3) of water for about two minutes.

2. Repeat, using sulphur dioxide and a second sample of water.

3. Mix two solutions, filter into the other boiling tube and observe.

Question

What is the equation for the reaction?

2 By reaction between thiosulphate and hydrogen ions in aqueous solution

Requirements

4·0 g sodium thiosulphate
6 cm^3 concentrated hydrochloric acid
500 cm^3 beaker
ignition tube
test tube
film-strip projector and fittings†

† See experiment 6.1A.

Procedure

1. Fill the ignition tube with crystals of sodium thiosulphate (c. 3·0 g), tip them into the test tube, add purified water, and shake to dissolve them.

2. Transfer the solution to the beaker and top up with purified water.

3. In darkened surroundings, arrange the film-strip projector and fittings, and pass the restricted beam of light through the contents of the beaker for steps *4* and *5*.

4. Add a quarter of a test tube (c. 5 cm^3) of concentrated hydrochloric acid to the contents of the beaker and stir with the test tube.

5. Observe continuously for at least five minutes, and again when at least fifteen minutes have elapsed.

Questions

1. What is the equation for the reaction?
2. How can the observations in step *5* be explained?

6.5 Preparation of colloids by arrest of molecular growth

A *Colloidal silver*

Requirements

4·0 g dextrin
3·0 g sodium hydroxide pellets
12 cm^3 industrial alcohol
12 cm^3 bench silver(I) nitrate solution
10 cm^3 bench dilute hydrochloric acid
two 100 cm^3 beakers
250 cm^3 beaker
test tube
boiling tube
teat pipette

Procedure

1. Dissolve a quarter of a test tube of dextrin (c. 3·0 g) in a 100 cm^3 beaker of water.

2. Dissolve twenty-four pellets of sodium hydroxide (c. 2·5 g) in the other 100 cm^3 beaker of water.

3. In the 250 cm^3 beaker, mix the two solutions and add half a test tube (c. 10 cm^3) of silver(I) nitrate solution.

4. Boil the mixture for about twenty minutes to complete the reduction of silver(I) ions to silver (test for this by teat pipette withdrawal of samples and subsequent addition of equal volumes of dilute hydrochloric acid).

5. Quarter fill a boiling tube (c. 15 cm^3) with the colloid, add an equal volume of the alcohol, shake and observe.

Questions

1. What are the functions of the dextrin?

2. What explanation can be given to the effect produced by the addition of alcohol in step *5*?

B *Colloidal silver(I) chromate*

Requirements

100 cm^2 sheet gelatine
20 cm^3 0·2M trisodium orthophosphate solution
20 cm^3 0·2M aluminium sulphate solution
25 cm^3 bench silver(I) nitrate solution
25 cm^3 bench potassium chromate solution
two 100 cm^3 beakers
filter funnel
test tube
glass stirring rod

Procedure

1. Boil half a beaker of water, and in the meantime measure out about 75 cm^2 of sheet gelatine (c. 1·0 g) and break it up into small pieces.

2. Stop heating, and add the gelatine to the hot water. Stir until the gelatine has dissolved.

3. Divide the solution equally between the two 100 cm^3 beakers. Then (*a*) to one portion add a test tube (c. 20 cm^3) of silver(I) nitrate solution; (*b*) to the other portion add a test tube of potassium chromate solution.

4. Stir both these solutions, allow them to stand for five minutes, and then mix and observe closely.

5. (*a*) Filter a sample of the product.

(*b*) Try to coagulate portions of the colloidal silver(I) chromate by addition of (*i*) the trisodium orthophosphate solution and (*ii*) the aluminium sulphate solution.

Questions

1. What is the equation for the reaction?

2. What is the function of the gelatine?

3. How can the observations from step *5* be explained?

C *Colloidal mercury(II) iodide*

Requirements

100 cm^2 sheet gelatine
10 cm^3 bench mercury(II) chloride solution
10 cm^3 bench potassium iodide solution
five 100 cm^3 beakers
test tube
two boiling tubes
glass stirring rod

Procedure

1. Add half a test tube (c. 10 cm^3) of mercury(II) chloride solution to half a beaker (c. 50 cm^3) of water.

2. Repeat step *1*, but use potassium iodide solution and the second beaker.

3. Mix together, in the third beaker, about half the total volume of the diluted solutions from steps *1* and *2*. Observe exactly what happens and retain the product as a comparative standard.

4. Make a gelatine solution in the fourth beaker.†

5. Put a test tube (c. 20 cm^3) of the gelatine solution and an equal volume of the diluted mercury(II) chloride solution from step *1* into a boiling tube, stir and allow to stand for about five minutes.

6. Repeat step *5*, employing the diluted potassium iodide solution from step *2*.

7. Pour the two solutions obtained in steps *5* and *6* into the fifth beaker. Observe and compare with the standard from step *3*.

† See experiment 6.5B.

Questions

1. What is the equation for the preparation?

2. What explanation may be given for the different colours of the precipitated and colloidal mercury(II) iodide?

6.6 Preparation of colloids by peptisation of a precipitate

A Prussian blue

Requirements

20 cm³ 0·2M trisodium orthophosphate solution
20 cm³ 0·2M aluminium sulphate solution
25 cm³ c. 5 per cent oxalic acid solution
10 cm³ bench iron(III) chloride solution
10 cm³ bench potassium hexacyanoferrate(II) (i.e. potassium ferrocyanide) solution
Buchner flask and funnel with attachments for suction filtration
two 100 cm³ beakers
test tube
boiling tube
film-strip projector and fittings†

Procedure

1. Mix half a test tube (c. 10 cm³) of iron(III) chloride solution with an equal volume of potassium hexacyanoferrate(II) solution in a boiling tube.

2. Filter off the precipitate on the Buchner funnel and wash the precipitate with three half test tube portions (c. 10 cm³ each) of purified water. Discard the filtrate and washings, wash out the Buchner flask and reassemble.

3. Pour half a boiling tube (c. 25 cm³) of the oxalic acid solution over the precipitate and allow to filter without suction. Transfer the filtrate back to the boiling tube.

4. Preferably in darkened surroundings, pass the restricted beam of light through the contents of the boiling tube.

5. Try to coagulate two different portions of the colloidal Prussian blue by adding trisodium orthophosphate solution to one of them and aluminium sulphate solution to the other.

† See experiment 6.1A.

Questions

1. What is the equation for the reaction?

2. What is the chemical name of Prussian blue?

B *Vanadium(V) oxide*

Requirements

1·2 g ammonium metavanadate
10 cm^3 bench dilute nitric acid
two 100 cm^3 beakers
two 250 cm^3 beakers
filter funnel
−10°C to 110°C thermometer
ignition tube
glass stirring rod
film-strip projector and fittings†

Procedure

1. Put half an ignition tube (c. 1·0 g) of ammonium metavanadate into a 100 cm^3 beaker and add half a test tube (c. 10 cm^3) of dilute nitric acid. Stir the mixture thoroughly until all the salt has decomposed.

2. Filter the contents of the beaker and wash the residue with purified water until the filtrate is distinctly coloured.

3. Puncture the filter paper and wash the residue into a 250 cm^3 beaker with purified water until it is two-thirds full.

4. Stir well with the thermometer, then gently warm the suspension to about 40°C and maintain this temperature for about half an hour.

5. Filter off a test tube of this warm solution into a 250 cm^3 beaker about half full of purified water.

6. Preferably in darkened surroundings, pass a restricted beam of light through the contents of the 250 cm^3 beaker.

Questions

1. What is the equation for the reaction between ammonium metavanadate and nitric acid?

2. What is the normal colour and solubility in water of vanadium (V) oxide?

† See experiment 6.1A.

3. Can a laboratory sample of vanadium(V) oxide be peptised by employing the same method? Try this practically!

6.7 Preparation of a colloid by dispersion

Requirements

100 cm^3 beaker about half full of purified water, standing in a 12·5 cm × 6·5 cm crystallising dish containing ice, two silver electrodes and a 20-volt source of D.C. (from any suitable LT variable voltage supply)

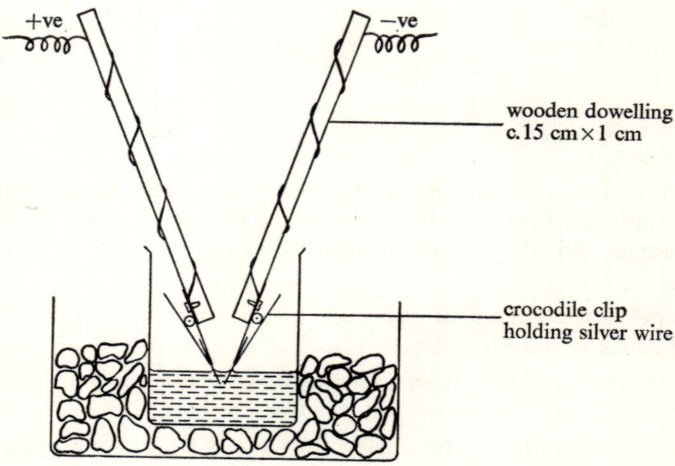

FIG. 11

Procedure

1. Clamp the electrodes so that the gap at the tips is about 0·2 cm.

2. Apply the current *with care*, thereby striking an arc, until the water is distinctly coloured.

Question

How can the characteristics of this colloid be practically investigated?

6.8 Dialysis of a mixture of starch and potassium iodide solution

Requirements

piece of thin cellophane (c. 15 cm × 15 cm)
1 cm^3 fresh sodium hypochlorite solution
6 cm^3 bench potassium iodide solution
30 cm^3 bench dilute hydrochloric acid
10 cm^3 starch indicator
250 cm^3 beaker
ignition tube
test tube
boiling tube
teat pipette

Procedure

1. Pour water into the beaker until the level is about 1·5 cm from the top, and dip the cellophane into it.

2. Place the piece of wetted cellophane over the top of the beaker and carefully press it down so that it makes good contact with the water.

3. Quarter fill the test tube (c. 5 cm^3) with starch indicator and add to it an equal volume of potassium iodide solution. Pour this mixture into the cellophane trough and set aside for ten minutes.

4. With the teat pipette put five drops of the sodium hypochlorite solution into a boiling tube and add dilute hydrochloric acid until it is half full.

5. Add an ignition tube (c. 4 cm^3) of the acidified sodium hypochlorite solution to the contents of the cellophane trough and observe.

6. Carefully remove and dispose of the cellophane trough and contents.

7. Add the same volume of the acidified sodium hypochlorite solution to the contents of the beaker and observe. Add an ignition tube of starch indicator and observe again.

Questions

1. What is the action of the sodium hypochlorite solution?

2. What explanation can be given for the observations in steps *5* and *7*?

3. What are the equations for the various reactions which have taken place?

6.9 Electrophoresis

Requirements

30 cm³ c. 2·5 per cent aquadag solution
filter funnel
12·5 cm × 1·5 cm U-tube with side arms standing in a 600 cm³ beaker of water, two platinum wire electrodes with mercury contacts and an 80-volt source of D.C. (from any conventional HT variable voltage supply as used for school experiments in Modern Physics)

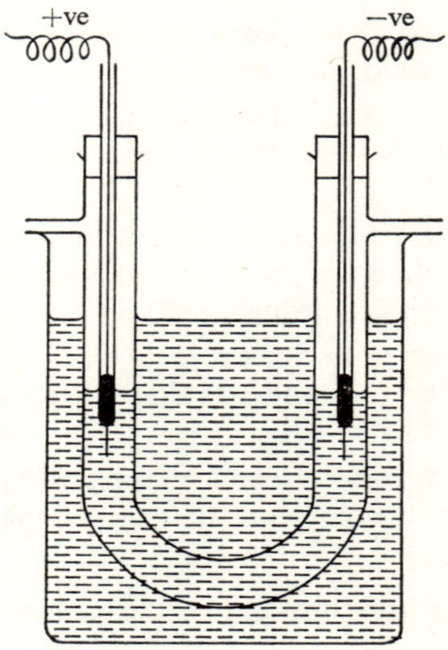

FIG. 12

Procedure

1. Quickly filter the aquadag solution through a coarse filter-paper into the U-tube.

2. Switch on the current *with care* and observe periodically.

Questions

1. What is aquadag?

2. What is the sign of the charge on the colloidal particles of aquadag?

.10 Preparation of gels

A Calcium carbonate gel

Requirements

5·0 g anhydrous calcium chloride
6·0 g hydrated sodium carbonate
1·0 g precipitated calcium carbonate
10 cm^3 bench dilute hydrochloric acid
20 cm^3 pipette
test tube
two boiling tubes, one with a bung to fit
two pieces of glazed paper (c. 8 cm × 8 cm)

Procedure

1. On separate pieces of glazed paper, weigh about 4·4 g of anhydrous calcium chloride and 5·7 g of sodium carbonate crystals.

2. Dissolve the salts in two separate half test tube portions (c. 10 cm^3) of water, each contained in a boiling tube.

3. (*a*) Carefully suck up most of the sodium carbonate solution into the 20 cm^3 pipette, and lower the tip of the pipette to the bottom of the boiling tube containing the calcium chloride solution.

(*b*) Slowly allow the sodium carbonate solution to run out, and at the same time gradually raise the pipette.

4. Insert the bung, invert the boiling tube and observe.

Questions

1. How can the final observation be explained?

2. What explanation may be given to the experimental comparison of the action of dilute hydrochloric acid on the gel and on precipitated calcium carbonate? Investigate practically.

B Alcohol gel

Requirements

15 g precipitated calcium carbonate
12 cm^3 glacial acetic acid
25 cm^3 industrial alcohol
100 cm^3 beaker
filter funnel
ignition tube
test tube
boiling tube
glass stirring rod

Procedure

1. Pour half a test tube (c. 10 cm^3) of glacial acetic acid into the 100 cm^3 beaker and add an equal volume of water.

2. Very slowly, stir in precipitated calcium carbonate until all signs of effervescence have ceased.

3. Filter the contents from step *2* through a previously moistened filter paper.

4. Put half an ignition tube (c. 2 cm^3) of the calcium acetate solution from step *3* into a boiling tube, followed by a test tube (c. 20 cm^3) of alcohol, and observe.

7

ADSORPTION

Introduction

Chromatography, an analytical method for the separation of various components of a mixture, can be divided into various types, all of which depend in some measure on adsorption. The system always contains a stationary phase, i.e. solid or liquid, to adsorb the mixture being separated, and a moving phase, i.e. liquid or gas, which passes through the stationary phase.

Chemicals which are being separated on, say, an activated aluminium oxide column are held on the aluminium oxide by forces which almost completely obey the Langmuir adsorption isotherm. When the adsorption is sufficiently strong to cause deviation from the isotherm, the elution zones tend to 'tail' because of the difficulty of desorption. During elution, the adsorbed solute fails to pass down the column in a sharp band and may lead to difficulty in effecting a good separation.

The extension of the techniques of chromatography to other fields, such as partition and ion exchange methods, merely uses different aspects of adsorption phenomena.

Questions

1. What are the various types of chromatography?

2. How does the Langmuir adsorption isotherm differ from the Freundlich adsorption isotherm?

3. What is the stationary phase in paper chromatography?

1 Investigation into the nature of screened methyl orange

Requirements

25 cm^3 c. 20 per cent industrial alcohol
60 cm^3 bench dilute hydrochloric acid
60 cm^3 bench sodium hydroxide solution
screened methyl orange indicator solution

250 cm^3 beaker
two boiling tubes
dropping pipette
stapler
hot air blower
three strips of white blotting paper (11 cm × 1 cm), each looped on a c. 15 cm length of stiff wire, as shown in figure 13:

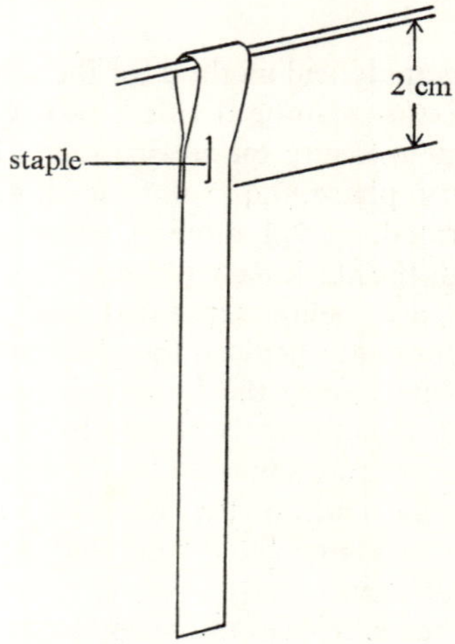

Fig. 13

Procedure

1. Pour the alcohol solution into the beaker.

2. Assemble the blotting paper strips on the wire. Using the dropping pipette, spot (c. 0·5 cm diameter) screened methyl orange on each strip about 2 cm from the unstapled end. Dry these spots over the hot air blower.

3. Rest the wire on the lip of the beaker with the ends of the blotting paper strips just dipping into the solvent. When the solvent fronts reach the bottom of the loops, remove the strips from the solvent.

4. (*a*) Develop the first strip by dipping it into a boiling tube of bench dilute hydrochloric acid.

(*b*) Develop the second strip by dipping it into a boiling tube of bench sodium hydroxide solution.

(*c*) Keep the third strip as a comparison.

Observations

Complete a table with the following headings:

COLOURS FOR:		
strip dipped in dilute hydrochloric acid	strip dipped in sodium hydroxide solution	comparison strip

Question

What is the nature of the screening?

7.2 Investigation into the acid and alkaline colours of the indicators present in B.D.H. universal indicator

Requirements

B.D.H. universal indicator solution
reagent bottle of 0·88 ammonia
reagent bottle of concentrated hydrochloric acid
two Green's No. 904 filter papers (11 cm diameter)
dropping pipette
plastic centimetre ruler
coloured pencils

Procedure

1. Using the dropping pipette, place B.D.H. universal indicator in the centre of an unfolded filter paper until the spot is no greater than 1 cm in diameter.

2. Wash out the dropping pipette and fill it up with water. Hold the filter paper in one hand and very carefully place a drop of water in the centre of the spot of indicator.

3. When the drop of water has been soaked up, add a further drop. Continue to add water in this manner until the diameter of the coloured area on the filter paper is about 6 cm. *Take great care not to flood the paper during this procedure.*

4. Develop the chromatogram by holding it over a bottle of concentrated ammonium hydroxide solution.

5. Using coloured pencils, immediately record the colours (which quickly fade) of the rings of indicators.

6. Prepare a second paper chromatogram by repeating steps *1* to *4*.

7. Develop the chromatogram prepared in step *6* by holding the paper over a bottle of concentrated hydrochloric acid.

8. Using the coloured pencils, again draw the coloured rings.

Questions

1. How many different indicators are used to make up B.D.H. universal indicator solution?

2. What are the acid and alkaline colours of each indicator?

3. What indicators would give the colour changes in question *2*?

7.3 Determination of the R_f values of the dyes present in coloured inks

Theory

In paper chromatography, for a given component and solvent, the ratio

$$\frac{\text{distance moved by centre of component spot}}{\text{distance moved by solvent front}}$$

is known as the R_f value and is a constant (both distances refer to the same base line).

Requirements

solvent mixture of 30 cm^3 butan-1-ol, 10 cm^3 industrial alcohol and 10 cm^3 c. 2M aqueous ammonium hydroxide
six coloured permanent inks (e.g. blue, blue-black, black, green, red and turquoise)
c. 7 cm long spotting rod, as illustrated on page 93
600 cm^3 beaker (tall form without spout)
glass plate (c. 13 cm $\times$ 13 cm)
Whatman No. 1 filter paper (15 cm $\times$ 15 cm)
clock or watch
piece of cardboard (c. 20 cm $\times$ 20 cm)
two clothes pegs
centimetre ruler
stapler
hot air blower

Procedure

1. In pencil, label the filter paper with six points, X, 2 cm apart and in a line 2 cm from one edge. Use the spotting rod to spot each X with a different coloured ink. *The spots must not have a diameter larger than 0·5 cm.* Dry the ink spots over the hot air blower and then staple the filter paper, as shown in figure 14.

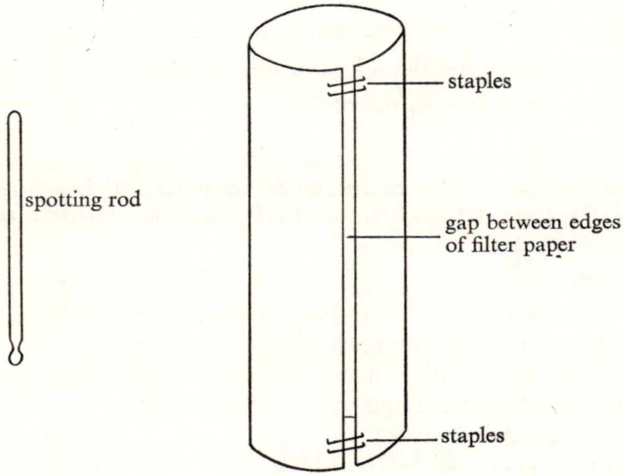

FIG. 14

2. Put the solvent mixture in the beaker. Simultaneously stand the cylindrically shaped filter paper in the solvent so that it does not touch the side of the beaker and note the time. Cover the beaker with the glass plate.

3. Allow the chromatograms to run for at least an hour (or until the solvent front approaches the top of the paper).

4. Remove the cylindrical filter paper from the beaker, carefully unstaple it and lay it on the piece of cardboard. Rule a pencil line along the solvent front.

5. Secure the filter paper in the clothes pegs and dry over the hot air blower.

6. Measure the distances involved in the table below.

Observations and Calculations

Tabulate for each ink thus:

colour of ink	colour of components	running time	cm moved by solvent	cm moved by component dye	R_f value

Questions

1. Which inks contain the same dyes?

2. Why in step *1* must the ink spot not be larger than 0·5 cm?

3. What do the letters R_f denote?

7.4 Determination of the order of adsorption of bismuth, copper(II), lead(II) and mercury(II) ions on filter paper*

Requirements

solvent mixture of 100 cm^3 acetone, 100 cm^3 2-methyl propan-2-ol and 50 cm^3 bench dilute nitric acid
separate solutions, in dilute nitric acid, of bismuth nitrate, copper(II) nitrate, lead(II) nitrate and mercury(II) nitrate
hydrogen sulphide generator
five pairs of glass plates (c. 13 cm × 13 cm), one of each pair with a 0·5 cm diameter hole drilled in the centre
five Whatman No. 1 filter papers (11 cm diameter)
five 100 cm^3 beakers
ignition tube
five dropping pipettes
pair of scissors
hot air blower

Procedure

1. Prepare a mixture of the four nitrates, using the dropping pipettes, by putting two drops of each solution into the ignition tube and mixing.

2. (*a*) Using the mixture of nitrates, spot the centre of a filter paper so that the spot is no larger than c. 0·5 cm. In pencil, label the filter paper at the edge.

(*b*) Repeat step *2* (*a*) separately for each nitrate solution.

3. Dry the spots.

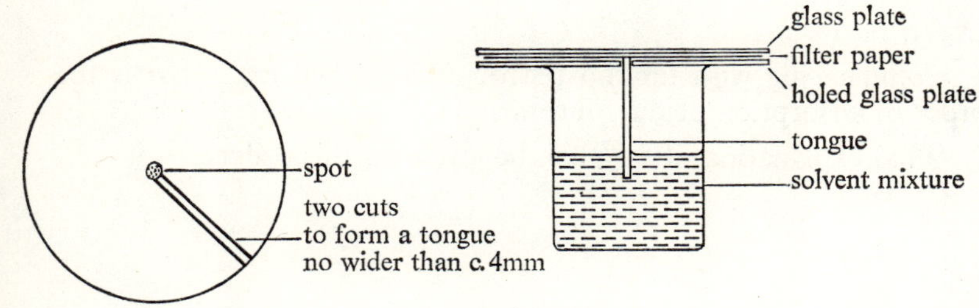

FIG. 15

4. Assemble the five sets of apparatus as shown in figure 15, with simultaneous insertion of the five tongues in the solvent mixture.
5. Allow each chromatogram to run for the same length of time (about ninety minutes, or until the solvent fronts almost reach the edges of the glass plates).
6. Remove the top plate and mark in pencil the position of each solvent front.
7. Develop each chromatogram in an atmosphere of hydrogen sulphide (play the jet from the hydrogen sulphide generator all over the top side of the paper).
8. Dry each paper and measure the distances travelled by (*a*) the solvent front and (*b*) the metallic ion(s), measured to the centre of the ring.

Observations and Calculations

Complete a table with the following headings:

metallic ion(s)	running time in min	developed colour of final ring(s)	cm moved by solvent	cm moved by metallic ion(s)	R_f value(s)
Bi^{3+}					
Cu^{2+}					
Pb^{2+}					
Hg^{2+}					
mixture					

Questions

1. Commencing with the most strongly adsorbed ion, what is the order of adsorption of the four ions?

2. What explanation, if any, can be given for this order?

7.5 Analysis of two solutions which may contain ions of cobalt(II), copper(II), iron(III) and nickel(II)*

Requirements

separate solutions, in dilute hydrochloric acid, of cobalt(II) chloride, copper(II) chloride, iron(III) chloride and nickel(II) chloride
two different solutions, each having some of the specified cations
solvent mixture of 86 cm^3 acetone, 6 cm^3 concentrated hydrochloric acid and 8 cm^3 purified water
saturated solution of rubeanic acid (dithio-oxamide) in ethyl alcohol contained in a plastic spray (hair lacquer type)
reagent bottle of 0·88 ammonia
vaseline
250 cm^3 beaker
gas jar (30 cm high × 7·5 cm diameter)
test tube
dropping pipette
glass plate (c. 9 cm × 9 cm)
two glass plates (c. 9 cm × 4·5 cm)
two wide elastic bands to stretch to c. 9 cm
two pieces Whatman No. 1 filter paper (30 cm × 6 cm)
two clothes pegs
hot air blower

Procedure

1. Pour the solvent mixture into the clean gas jar.

2. Lightly vaseline the top of the gas jar and cover with the two smaller glass plates. Place the square piece of glass on top and secure these to the gas jar by means of two elastic bands as shown in figure 16 on page 97.

3. Swirl the liquid round the lower sides of the gas jar and allow it to stand for half an hour.

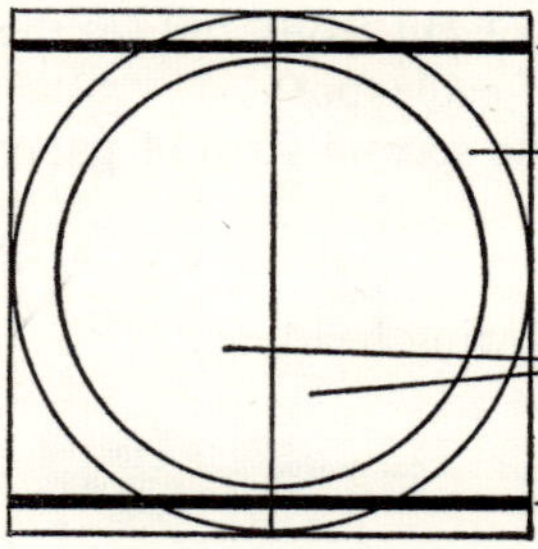

FIG. 16

4. In pencil, rule a line 2 cm from one end, across both strips of filter paper, and on each of these lines mark and label points for application of three spots.

5. Using a clean dropping pipette for each solution, apply spots of the four known and the two unknown solutions to the two strips of paper in their appropriate places, ensuring that the spots are *no larger than 0·5 cm in diameter.*

6. Dry the papers over the hot air blower.

7. After equilibration has occurred in step *3*, carefully remove the elastic bands and the top glass plate, and slide open the lower glass plates about 0·1 cm apart.

8. Quickly insert one of the filter paper strips until the lower edge of the paper just touches the surface of the solvent. Close the glass plates, bend over the projecting strip of filter paper, replace the top glass plate and elastic bands, and leave undisturbed for one hour.

9. Ensuring minimum loss of solvent vapour from the gas jar, withdraw the paper and hold each end in a clothes peg. Dry the paper over the hot air blower.

10. (*a*) Suspend the paper vertically and spray it with the alcoholic rubeanic acid.

(*b*) Pass the paper horizontally over a 250 cm^3 beaker containing a test tube (c. 20 cm^3) of 0·88 ammonia.

11. Dry the paper again and ring the spots in pencil. Measure the distances travelled by (*a*) the solvent front and (*b*) the metallic ions (to the most prominent part of each spot).
12. Repeat steps *8* to *11* for the second strip of paper.

Observations and Calculations

Complete a table with the following headings:

metallic ion(s)	running time in min	colour of final spot(s)	cm moved by solvent	cm moved by metallic ion(s)	R_f value(s)
Co^{2+}					
Cu^{2+}					
Fe^{3+}					
Ni^{2+}					
unknown solution I					
unknown solution II					

Questions

1. What are the ions present in the unknown solutions?
2. What is the structural formula for rubeanic acid?

7.6 Separation of coloured cations on a thin film of activated aluminium oxide*

Requirements

11 g aluminium oxide for chromatographic analysis
5 cm^3 aqueous solution containing copper(II), iron(II) and cobalt(II) ions
test tube
dropping pipette
two glass (or perspex) plates (12 cm × 12 cm), one with a 0·5 cm diameter hole drilled in the centre
coloured pencils

Procedure

1. Pour half a test tube of the aluminium oxide (c. 9·0 g) into the centre of the non-drilled glass plate.

2. Place the second glass plate on top of the first, and lightly move it in a circular motion until the layer of aluminium oxide is evenly distributed (diameter c. 10 cm, thickness c. 0·1 cm).

3. Using the dropping pipette, slowly drip the solution of the metal ions through the central hole. Continue adding the solution until either three coloured bands can be clearly seen or the solvent front reaches the periphery of the aluminium oxide.

4. Using coloured pencils, record the chromatogram.

Questions

1. What is the order of adsorption of the coloured ions on the aluminium oxide?

2. How can this experiment be adapted to obtain complete separation of the three metal ions?

7 Preparation of a column for adsorption chromatography

Requirements

25 cm^3 burette (pinch-clip type) and a plug of glass wool, with an appropriate adsorbent and eluant
100 cm^3 beaker
dropping pipette
glass stirring rod (at least as long as the column)

Procedure

1. Prepare a slurry of the solvent and adsorbent by stirring the mixture thoroughly in the 100 cm^3 beaker to remove all air bubbles.

2. Pour about a third of a test tube (c. 7 cm^3) of solvent into the closed chromatography tube.

3. Using the glass rod, push a small plug of glass wool down through the solvent as far as the constriction in the tube (*do not push it any further or it will be difficult to remove*) and ensure there are no trapped air bubbles.

4. Pour the slurry into the tube and, if necessary, rinse it in with a little more solvent; *do not introduce air bubbles* (if they appear, the column must be prepared again).

5. Run out the solvent until the top of the column is covered by only a thin layer of solvent (c. 0·2 cm). *Do not let the top of the column become dry during the subsequent operations.*
6. Using the dropping pipette (to avoid disturbance of the surface of the column), very carefully add the experimental mixture to be chromatogrammed.
7. (*a*) Allow the sample from step *6* to run into the column. When the sample is just about to disappear into the column, add (again by dropping pipette) about 1 cm^3 of eluant.

(*b*) Watch for this 1 cm^3 of eluant to almost disappear through the surface, and add very carefully (initially by dropping pipette) eluant to fill up the tube prior to elution, as illustrated in figure 17.

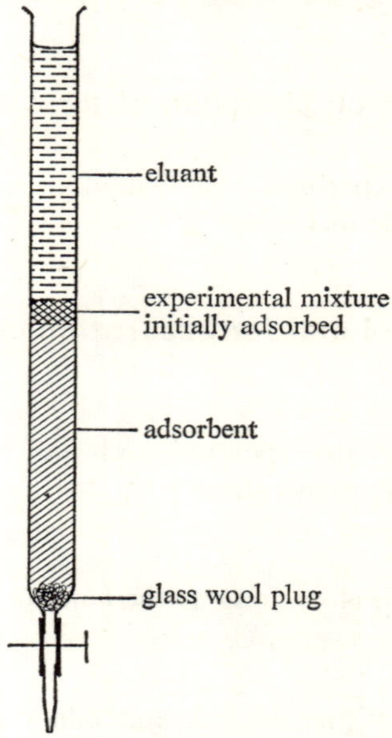

FIG. 17

7.8 Separation of selected anions and cations by using an ion exchange resin, Zeo-carb 225

Requirements

12 g Zeo-carb 225
150 cm^3 c. 4M hydrochloric acid
bench reagents: barium chloride, copper(II) sulphate, iron(III) chloride, potassium thiocyanate and silver(I) nitrate solutions, dilute hydrochloric and nitric acids, 0·88 ammonia

100 cm^3 beaker
two ignition tubes
two test tubes
apparatus for the preparation of a column (see experiment 7.7)

Procedure

1. Prepare a slurry using a test tube of Zeo-carb 225 (c. 10 g) and a test tube (c. 20 cm^3) of c. 4M hydrochloric acid.

2. Prepare the column as in experiment 7.7, steps *2* to *5*, using the acid in place of the solvent.

3. Run purified water through the column until the eluate gives a negative chloride test: the surface of the column must be left covered with a thin layer of water.

4. Fill an ignition tube with approximately equal volumes (c. 2 cm^3 each) of iron(III) chloride and copper(II) sulphate solutions, and mix well.

5. Employing the experimental mixture just prepared, carry out steps *6* and *7* of the last experiment with purified water as the eluant.

6. (*a*) Collect a test tube of eluate at a rate no faster than one drop per second. Qualitatively test for all the ions which may be present in the eluate.

(*b*) Continue passing purified water through the column until the tests for all the ions are negative.

7. (*a*) Carefully fill up the column with 4M hydrochloric acid and slowly elute the column until the eluate is coloured. Collect a test tube of the coloured eluate and test (omit the chloride test) for the presence of the original ions.

(*b*) Continue eluting the column until the tests prove negative.

8. Pour out the resin into a stock bottle, containing 4M hydrochloric acid, for future use.

Questions

1. What ions were found in steps *6* (*a*) and *7* (*a*)?

2. Is Zeo-carb 225 an anion or cation exchange resin?

3. Why was the slurry of Zeo-carb 225 prepared in step *1* made in 4M hydrochloric acid?

4. Why must the column never be allowed to dry out?

5. Why must the column be eluted very slowly?

6. What is Zeo-carb 225?

7.9 Separation of the dyes in blue-black ink on calcium carbonate

Requirements

5·0 g pure precipitated calcium carbonate
blue-black ink
test tube
apparatus for the preparation of a column (see experiment 7.7)
filter pump with rubber connecting tube for attachment to the burette jet

Procedure

1. Prepare a slurry using a test tube of the calcium carbonate (c. 4·0 g) and a test tube (c. 20 cm^3) of purified water.

2. Prepare the column as in experiment 7.7, steps *2* to *5*, using the filter pump to hasten the packing and the flow of water through the column.

3. Carry out steps *6* and *7* of experiment 7.7 with ink as the specimen, and purified water as the eluant.

4. Develop the column by sucking through purified water for about fifteen minutes.

Question

What is (*a*) the order and (*b*) the colour of the bands?

7.10 Purification of a sample of technical anthracene

Theory

Technical anthracene contains carbazole and naphthacene as impurities. When it is viewed under ultra-violet light, anthracene itself produces a blue-violet fluorescence, carbazole a deep blue fluorescence, and naphthacene a yellow colouration.

Requirements

0·3 g technical anthracene
25 g aluminium oxide for chromatographic analysis
200 cm^3 n-hexane
100 cm^3 beaker

250 cm^3 beaker
ignition tube
test tube
boiling tube
two pieces of white blotting paper (c. 8 cm × 8 cm) or filter paper
ultra-violet light source (DANGER TO EYES: DO NOT VIEW DIRECTLY)
apparatus for the preparation of a column (see experiment 7.7).
melting-point apparatus, including a 0°C to 360°C thermometer and two melting-point tubes

Procedure

1. In the 100 cm^3 beaker, dissolve a quarter of an ignition tube (c. 0·25 g) of technical anthracene in the minimum volume of n-hexane.

2. Prepare a slurry using a test tube of activated aluminium oxide (c. 20 g) and a test tube (c. 20 cm^3) of n-hexane.

3. Prepare the column as in experiment 7.7, steps *2* to *7*, employing the anthracene solution from step *1* above where necessary and n-hexane as eluant.

4. Elute, while viewing the column under ultra-violet light, with n-hexane. When the blue-violet fluorescent zone starts to percolate from the column, collect a sample of it in a clean boiling tube.

5. (*a*) Almost fill the 250 cm^3 beaker with water, heat to boiling point, and remove the source of heat.

(*b*) Stand the anthracene sample in the hot water and allow the solvent to evaporate completely.

(*c*) Dissolve the residual solid in the minimum volume of hot n-hexane (do not use a naked flame for heating it) and set aside to crystallise.

(*d*) Dry a few crystals between squares of blotting paper and determine their melting point.

6. Determine the melting point of the technical anthracene.

Questions

1. What is the accepted melting point of pure anthracene, and how does it compare with the technical and chromatogrammed specimens?

2. What are the formulae of anthracene, naphthacene and carbazole?

7.11 Separation of the pigments in grass

A Using a column of silica gel

Requirements

15 g silica gel for chromatographic analysis
handful of fresh dry grass
25 cm^3 acetone
75 cm^3 benzene (TOXIC:CARE)
mortar and pestle
250 cm^3 beaker
filter funnel
test tube
boiling tube
chip of porous pot
coloured pencils
apparatus for the preparation of a column (see experiment 7.7)

Procedure

1. Press a handful of grass into a mortar and add a test tube (c. 20 cm^3) of acetone. Carefully crush the grass until the solvent is a bright green.

2. Filter off the acetone solution into the boiling tube and add the chip of porous pot.

3. Heat the 250 cm^3 beaker of water until bubbles start to rise. Turn out the flame and stand the acetone solution in the hot water (even though heating does tend to denaturise the extract) until all the acetone has evaporated.

4. Remove the boiling tube from the hot water and allow to cool.

5. Meanwhile, prepare a slurry using a test tube of the silica gel (c. 12 g) and a test tube (c. 20 cm^3) of benzene.

6. Dissolve the grass extract in the boiling tube, using the *minimum* volume of benzene (since grass contains water, there may be two layers, the lower one being an aqueous layer).

7. Prepare the column as in experiment 7.7, steps *2* to *7*, employing the grass extract in step *6* and benzene as the eluant in steps *7* (*a*) and (*b*).

8. Elute with benzene until the grass pigments become separate coloured bands.

9. Record, as a coloured diagram, the positions of the coloured bands.

Questions

1. What are the various pigments extracted from the grass?

2. What is their order on the column?

3. Why should a highly concentrated grass extract be run into the column?

4. Why should great care be taken to prevent the top of the column from being disturbed?

B *Using a column of aluminium oxide*

Requirements

the same as for part A of this experiment, with the addition of (*a*) c. 25 g aluminium oxide for chromatographic analysis in place of the silica gel, (*b*) 75 cm^3 ethyl acetate, (*c*) 75 cm^3 dry cyclohexane in place of the benzene and (*d*) two extra, clean, dry boiling tubes and chips of porous pot

Procedure

1. Weigh two dry boiling tubes, each containing a chip of porous pot.

2. Prepare a slurry using a test tube of aluminium oxide (c. 20 g), a test tube (c. 20 cm^3) of the dry cyclohexane and a column, once again using a grass extract as described in part A.

3. Elute the column with the dry cyclohexane until the eluate is clear. Retain all the eluate in one of the dry weighed boiling tubes.

4. Now fill up the column with ethyl acetate and again elute the column as in step *3*, with separate collection of the eluate.

5. Boil a 250 cm^3 beaker of water and turn out the Bunsen burner.

6. Evaporate the eluants from their eluates by standing the boiling tubes in the hot water.

7. Remove the boiling tubes when all the eluants have been evaporated; dry, allow them to cool and reweigh.

Questions

1. What is present in each fraction?

2. What is the ratio by weight of the residues from the two fractions?

7.12 Separation of selected vapours on detergent powder

Requirements

20 g oven-dried detergent powder
four corked ignition tubes, each containing c. 3 cm^3 of one of the following: carbon disulphide, diethyl ether, acetone, and a mixture of equal volumes of all three
1 cm^3 syringe with needle
4 cm^3 porcelain crucible
black card (c. 10 cm × 10 cm)
gas chromatography column made from c. 100 cm of c. 0·4 cm internal diameter glass tubing and a blowpipe connected by thin rubber tubing to a coal gas supply

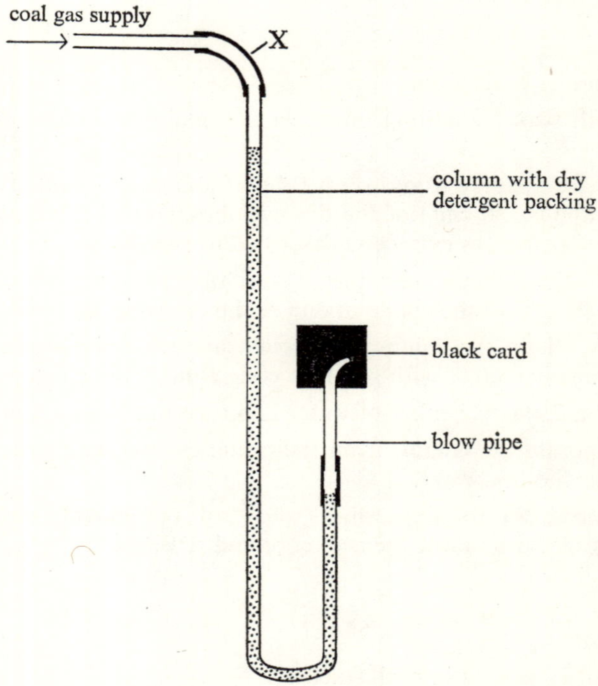

FIG. 18

Procedure

1. Add the dry detergent powder to the column; ensure even packing by addition of small portions and regular light tapping.

2. Attach the blowpipe and gas supply tubing to the column.

3. Pass the coal gas supply through the column and light at the jet.

4. After about five minutes, use the syringe to inject angularly into the column at X about 0·1 cm^3 of the volatile mixture.

5. Observe the ensuing flame colourations.

6. Determine the flame colourations of the three volatile liquids by their separate combustion in the crucible.

Question

What is the order of adsorption of the vapours in the column, and what explanation can be given for it?

7.13 Investigation of the adsorption of oxalic acid on activated charcoal and of the validity of the Freundlich adsorption isotherm*

Theory

The adsorption on activated charcoal of oxalic acid from its aqueous solution obeys the Freundlich adsorption isotherm:

$$\frac{x}{m}=Kc^n$$

where

n and K are constants for a given system at a given temperature,
x is the number of grams of oxalic acid adsorbed by
m g of activated charcoal, and
c is the concentration in mole per litre of oxalic acid at equilibrium.

The isotherm may also be expressed as

$$\log_{10}x-\log_{10}m=\log_{10}K+n\log_{10}C$$

and a graph of $\log_{10}x$ against $\log_{10}c$ should give a straight line of slope n.

Requirements

18 g activated charcoal
400 cm^3 0·125M oxalic acid solution
600 cm^3 c. 0·1M sodium hydroxide solution
phenolthymolphthalein indicator solution
seven dry 100 cm^3 beakers
specified volumetric ware plus two burettes, seven dry 250 cm^3 conical flasks and a 10 cm^3 pipette
clock or watch
seven squares of glazed paper (c. 8 cm × 8 cm)
fourteen adhesive labels

Procedure

1. Standardise the sodium hydroxide solution by titrating 10 cm^3 pipetted portions of the 0·125M oxalic acid solution with the alkali, using phenolthymolphthalein as indicator.

2. Number pairs of dry conical flasks and beakers from 1 to 7.

3. Weigh out exactly 2·00 g of the activated charcoal on a square of glazed paper creased lengthways, and shoot it into the first conical flask. Similarly add 2·00 g portions to the other six flasks.

4. Using burettes, add the liquid contents of the flasks according to the following table, noting the mean time of addition for each flask:

number of flask	cm^3 of oxalic acid	cm^3 of water
1	100	—
2	80	20
3	60	40
4	40	60
5	20	80
6	10	90
7	—	100

5. Shake each flask regularly over a period of twenty minutes.

6. Allow each conical flask to stand for a further ten minutes and then filter each oxalic acid solution into the correspondingly numbered beaker.

7. Titrate 10 cm^3 pipetted portions of oxalic acid from each beaker with the standardised sodium hydroxide solution, using phenolthymolphthalein as indicator, to obtain sets of concordant results.

Observations and Calculations

1. Calculate the molarity of the sodium hydroxide solution.

2. Prepare a table with the following headings:

number of flask	cm^3 of alkali required in step *1*	cm^3 of alkali required in step 7 less that required in flask number 7	mole/l of oxalic acid at equilibrium	g of oxalic acid absorbed by 2·00 g charcoal from 10 cm^3 of solution	g of oxalic acid absorbed by 1 g of charcoal	$\log_{10}c$	$\log_{10}x$
			c		x		

3. Plot a graph of x against c and comment on its form.

4. Plot a graph of $\log_{10}x$ against $\log_{10}c$ and from this graph find the values of n and K.

Questions

1. What is activated charcoal?

2. Why are the contents of flask number 7 titrated?

3. What are likely sources of error in this experiment, and how might they be overcome?

8

CATALYSIS

Introduction

Catalysts have usually been considered to affect a chemical change in one of two ways, depending on the state in which they exist with respect to the state of the reactants. Gaseous reactions may be carried out at a catalytic surface, i.e. heterogeneously, whereas ionic reactions in aqueous media may be catalysed by ions, i.e. homogeneously.

In heterogeneous catalysis, the catalyst surface provides a suitable arrangement of reacting molecules. Subsequent weakening and breaking of bonds culminates in effective yet less energetic collision and thus the reaction path is facilitated.

The potential energy curves for an uncatalysed and heterogeneously catalysed exothermic reaction are shown in figure 19.

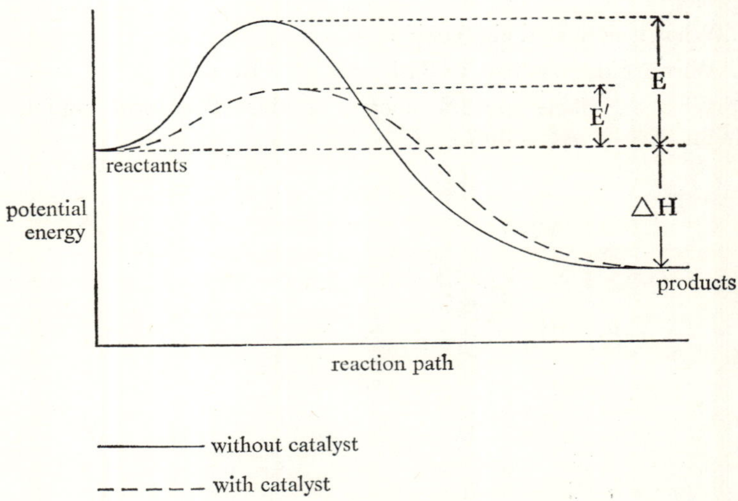

FIG. 19

The effect of a positive catalyst is to lower the barrier due to the energy of activation from E to E′. This energy barrier still has to be surmounted for the reaction to take place.

When a positive catalyst and the species present in a chemical change are homogeneous, the catalyst may react with a reactant to produce a relatively unstable intermediate compound which then breaks down to yield the original products. The catalyst provides an alternative route with a lower activation energy via the valley of the intermediate compound, as shown in figure 20.

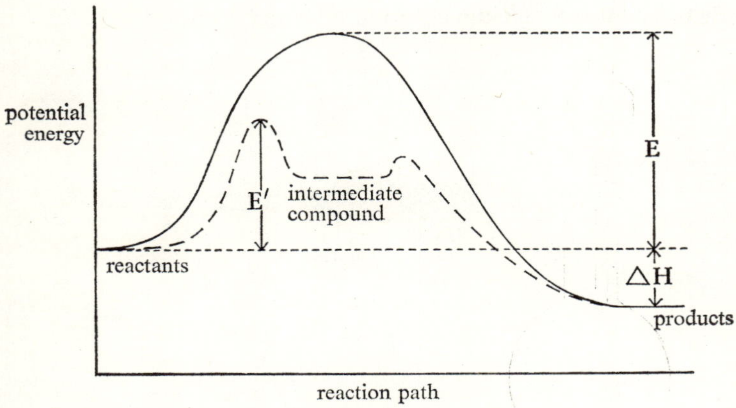

FIG. 20

The energy of a reaction, △H, is unchanged by catalysis, whether it is heterogeneous or homogeneous, although the reaction pathway is altered. Furthermore, bond making and breaking is involved in surface catalysis as well as in the formation of intermediate compounds. These universal features of catalysis support the latest views that there is only one mode of catalysis: catalysts are reactants which are chemically regenerated in the reaction system!

Question

What is the significance of the second peak on the catalysed reaction pathway in figure 20?

8.1 Combustion of carbon monoxide in air catalysed by water

Requirements

7.0 g fused calcium chloride
10 cm^3 formic acid
50 cm^3 concentrated sulphuric acid
two 1000 cm^3 round-bottomed flasks (one fitted with a bung)
boiling tube
carbon monoxide generator† consisting of a 250 cm^3 round-bottomed flask fitted with a 100 cm^3 dropping funnel, a horizontal delivery tube, a calcium chloride tube and a blowpipe

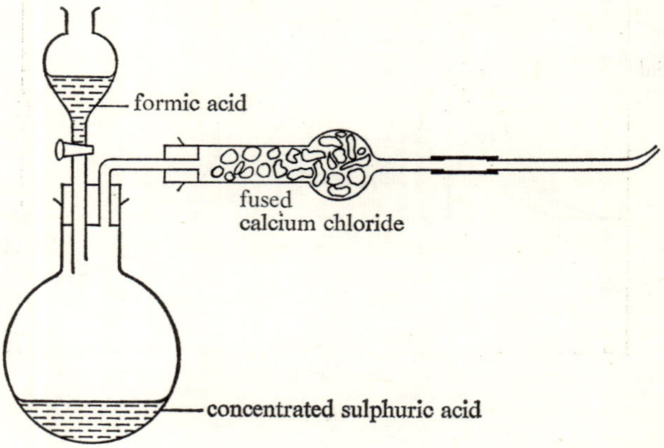

FIG. 21

Procedure

1. Remove the bung from the 1000 cm^3 flask, add half a boiling tube (c. 25 cm^3) of concentrated sulphuric acid, replace the bung and swill the acid round the flask (not in the neck) to dry the enclosed air.

2. Meanwhile, add formic acid dropwise to the concentrated sulphuric acid in the carbon monoxide generator flask.

3. When all the air has been expelled from the generating apparatus (after about two minutes of steady evolution of gas), light the jet and remove the bung from the flask of dry air.

† The generator is best left assembled if successive performances of the combustion are to be carried out.

4. Slide the flask of dry air over the flame, observe, and then withdraw the flask.

5. Insert the carbon monoxide flame into the flask of undried air and observe.

Questions

1. What is the equation for the reaction between concentrated sulphuric acid and formic acid?

2. Why should the formic acid be added to the concentrated sulphuric acid and not vice versa?

3. Although the carbon monoxide is generated by employing concentrated sulphuric acid, why is a calcium chloride tube essential?

.2 Oxidation of methanol catalysed by platinum

Requirements

10 cm^3 methanol
15 cm length of platinum wire with one end sealed in a 15 cm glass tube
250 cm^3 conical flask
piece of cardboard (c. 10 cm × 10 cm)

Procedure

1. Make a spiral by winding the platinum wire round a pencil and then withdrawing the pencil.

2. Pour 10 cm^3 of methanol into the conical flask.

3. With a small Bunsen flame, just slightly warm the alcohol.

4. With a non-luminous flame, heat the spiral to red heat and immediately lower it into the conical flask so that the spiral does not touch the alcohol. (Have the piece of cardboard ready to cover the mouth of the beaker in case the alcohol catches fire.)

5. Observe and smell the vapours in the beaker.

Questions

1. What is the equation for the reaction which has taken place?

2. What other catalysts may be employed for the oxidation of this alcohol? Consult a textbook of organic chemistry.

8.3 Oxidation of ammonia catalysed by platinum*

Requirements

10 cm^3 freshly prepared iron(II) sulphate solution (made up in cold dilute sulphuric acid)
1 cm^3 0·88 ammonia
potassium iodide/starch paper
'flag' made of platinum foil and wire, supported in a glass tube c. 25 cm long
oxygen cylinder (or apparatus for laboratory preparation of oxygen)
500 cm^3 round-bottomed flask fitted with a bung
dropping pipette
sheet of white card (c. 25 cm × 25 cm)
test tube

Procedure

1. By downward delivery, fill the flask with oxygen.

2. From the dropping pipette, add about three drops of 0·88 ammonia and immediately close the neck with the bung.

3. Rotate the flask several times and then clamp it in a horizontal position in front of the white card.

4. Remove the bung from the flask and rapidly (*a*) place a non-luminous Bunsen flame near the mouth of the flask, (*b*) make the platinum foil red hot, and (*c*) simultaneously thrust the red-hot foil into the flask and remove the Bunsen flame.

5. After all signs of reaction have ceased, add a test tube (c. 20 cm^3) of water to the flask and agitate.

6. Divide the aqueous solution into two portions. To one add a piece of potassium iodide/starch paper, to the other add about 10 cm^3 of the acidified iron(II) sulphate solution, and observe.

Question

What are the equations for all the reactions which have taken place?

8.4 Hydrolysis of sucrose catalysed by hydrogen ions

Requirements

4·0 g sucrose
6 cm^3 Fehling's solution A
6 cm^3 Fehling's solution B

2 cm^3 bench dilute hydrochloric acid
250 cm^3 beaker
ignition tube
test tube
boiling tube

Procedure

1. Wash out the various tubes with purified water to ensure that they are acid free.

2. Put a quarter of a test tube (c. 3·5 g) of sucrose and a test tube (c. 20 cm^3) of purified water into a clean boiling tube, and warm to dissolve the sucrose. By employing the second clean boiling tube, divide the sucrose solution into two approximately equal volumes.

3. Add a quarter of an ignition tube (c. 1 cm^3) of dilute hydrochloric acid to the contents of one boiling tube.

4. Quarter fill the test tube (c. 5 cm^3) with Fehling's solution A, add about an equal volume of Fehling's solution B, and shake well to mix. Divide this Fehling's solution equally between the contents of the two boiling tubes.

5. Simultaneously heat both boiling tubes in the 250 cm^3 beaker of boiling water.

Questions

1. What is Fehling's solution, and what property of the hydrolysis products does it detect?

2. What is the equation for the hydrolysis of sucrose, and what are the names of the products?

5 Decomposition of hypochlorite ions catalysed by transition metal ions

Requirements

50 cm^3 sodium hypochlorite solution
1 cm^3 of bench solutions containing the following ions: chromium(III), manganese(II), iron(II), iron(III), cobalt(II), nickel(II), copper(II) and zinc
500 cm^3 beaker
eight test tubes
eight adhesive labels

Procedure

1. Quarter fill each test tube (c. 5 cm^3) with sodium hypochlorite solution.

2. To each test tube, appropriately labelled, add a quarter of an ignition tube (c. 1 cm^3) of a different transition ion solution.

3. Stand the test tubes in the beaker and add water to a depth of about 5 cm.

4. Gently warm the contents of the beaker and test each tube regularly by insertion of a glowing splint.

Questions

1. What transition metal ions are suitable for the decomposition of hypochlorite ions, and what is their order of effectiveness?

2. What are the ionic equations for (*a*) the catalytic decomposition and (*b*) the atmospheric decomposition of hypochlorite ions?

8.6 Reaction between copper and nitric acid catalysed by nitrous acid

Requirements

2·0 g copper turnings
0·2 g sodium nitrite
1 cm^3 20-volume hydrogen peroxide
15 cm^3 30 per cent nitric acid
three test tubes

Procedure

1. Cover a few copper turnings in each of the three test tubes with about 2·5 cm of the nitric acid.

2. Immediately add a pinch of sodium nitrite to one test tube and upon the steady evolution of gas, add a little hydrogen peroxide to one of the other two tubes and shake.

3. Observe all three tubes for several minutes.

Questions

1. What explanations can be given for the observations?

2. What are the equations for the reactions which have occurred?

3. Why is there hardly any action between pure aluminium and nitric acid under any conditions?

7 Investigation of the catalytic action of silver(I) ions on the reaction of persulphate ions with chromium(III) and manganese(II) ions

Requirements

0·5 g chromium(III) sulphate crystals
0·5 g manganese(II) sulphate crystals
5·0 g ammonium persulphate
1 cm^3 20-volume hydrogen peroxide
5 cm^3 bench silver(I) nitrate solution
two 250 cm^3 conical flasks
filter funnel
test tube
boiling tube

Procedure

1. Put two boiling tubes (c. 110 cm^3) of water into a conical flask, add a pinch each of chromium(III) sulphate and manganese(II) sulphate, and boil.

2. While keeping the mixture boiling, add, over an interval of five minutes, three successive portions of ammonium persulphate, each about the size of a salt-spoonful.

3. Boil until all signs of reaction have ceased, filter the reaction mixture into the second conical flask, and retain both the residue and the filtrate.

4. Add drops of hydrogen peroxide to the filtrate from step *3* and observe.

5. Wash the residue from step *3* back into the first conical flask, using a boiling tube (c. 55 cm^3) of water, add a quarter of a test tube (c. 5 cm^3) of silver(I) nitrate solution, and boil. Repeat step *2* and subsequently filter, with retention and observation of the filtrate.

Questions

1. What are the ionic equations for the reactions which have taken place?

2. What may be inferred about the oxidation-reduction systems involved in the experiment?

3. By repeating the experiment without the presence of chromium(III) ions at the outset, what light may be thrown on the theories of catalysis?

8.8 Reaction between permanganate and oxalate ions auto-catalysed by manganese(II) ions*

Theory

Manganese(II) ions are one of the products of the reaction between permanganate and oxalate ions in acid solution:

$$2MnO_4^-(aq) + 5(COO)_2^{2-}(aq) + 16H^+(aq) \longrightarrow 10CO_2(g) + 2Mn^{2+}(aq) + 8H_2O(l)$$

and these manganese(II) ions catalyse the reaction.

Requirements

50 cm^3 saturated oxalic acid solution
5 cm^3 saturated manganese(II) sulphate solution
100 cm^3 c. 0·02M potassium permanganate solution
100 cm^3 bench dilute sulphuric acid
two 500 cm^3 beakers
stop-clock
−10°C to 110°C thermometer
two glass stirring rods
ignition tube
test tube
two boiling tubes

Procedure

1. Two-thirds fill each beaker with water and warm to about 40°C.

2. To each beaker add a test tube (c. 20 cm^3) of the oxalic acid solution and two test tubes (c. 40 cm^3) of the dilute sulphuric acid, and to one of the beakers add an ignition tube (c. 4 cm^3) of the manganese(II) sulphate solution.

3. Thoroughly stir the contents of each beaker.

4. Pour two test tubes (c. 40 cm^3) of the potassium permanganate solution into each of the two boiling tubes.

5. Simultaneously add a portion of the permanganate solution to each of the two beakers and start the clock.

6. With constant stirring, allow the contents of the two beakers to stand; observe and time appropriately.

Question

What is the relative efficiency of the manganese(II) ions as a catalyst?

9 Bromination of acetone autocatalysed by bromide ions

Theory

Hydrogen bromide is one of the products of the reaction between bromine and acetone:

$$CH_3COCH_3(l) + Br_2\,(CCl_4) \longrightarrow CH_3COCH_2Br(CCl_4) + HBr(CCl_4)$$

and acts as a catalyst for the reaction.

Requirements

6 cm^3 c. 5 per cent solution of bromine in carbon tetrachloride
25 cm^3 acetone
250 cm^3 beaker
burette
stop-clock
−10°C to 110°C thermometer
test tube
boiling tube
c. 500 cm^3 bench sodium hydroxide solution contained in a stoppered Winchester 'quart' bottle for residues†

Procedure

1. Half fill the beaker with water and warm to about 50°C.

2. Pour a test tube (c. 20 cm^3) of acetone into the boiling tube.

3. From the burette, add 1·0 cm^3 of the bromine solution and swirl. Immediately stand the reaction mixture in the hot water and start the clock. Observe and record the time for the disappearance of bromine.

4. Cool the contents of the boiling tube in a stream of cold tap water.

5. Add a further 1·0 cm^3 of the bromine solution, note the time of addition, and record the time interval for the loss of colour.

6. Repeat step *5* until a total of 5·0 cm^3 of the bromine solution has been added.

7. Immediately dispose of the reaction mixture into the Winchester 'quart' bottle, shake and stopper. (The hydrogen bromide is neutralised and the lachrymatory products are contained in an enclosed volume.)

† When the experiment is performed individually by members of a class, employ this residue bottle communally.

Questions

1. Does the experimental evidence support the autocatalytic action of hydrogen bromide?

2. What further reaction may be performed to confirm this evidence?

8.10 Titrimetric investigation of the catalytic effect of copper(II) ions on the reaction between dichromate and iodide ions in acid solution*

Theory

The reaction between potassium dichromate and potassium iodide in aqueous acid solution produces iodine:

$$Cr_2O_7^{2-}(aq) + 6I^-(aq) + 14H^+(aq) \rightarrow 2Cr^{3+}(aq) + 7H_2O(l) + 3I_2(aq)$$

and is catalysed by copper(II) ions. By recording the times for the reappearance of iodine after treatment of the reaction mixture with successive small, equal volumes of sodium thiosulphate solution, the reaction may be followed quantitatively.

Requirements

150 cm³ 0·008M potassium dichromate solution
45 cm³ 0·1M sodium thiosulphate solution
20 cm³ 0·001M copper(II) sulphate solution
180 cm³ c. 6 per cent potassium iodide solution
20 cm³ glacial acetic acid
40 cm³ starch indicator
specified volumetric ware plus a 25 cm³ pipette and a 250 cm³ conical flask
stop-clock
test tube

Procedure

1. Pipette 20 cm³ of 0·008M potassium dichromate solution into a conical flask.

2. From separate burettes†, add (*a*) 2·5 cm³ of glacial acetic acid and (*b*) 1·0 cm³ of 0·1M sodium thiosulphate solution to the potassium dichromate solution.

† When the experiment is performed individually by a class, it is best to have several communal burettes.

3. Add one-third of a test tube (c. 7 cm^3) of starch indicator, and then stand the reaction flask on a white tile.

4. Pipette 25 cm^3 of the potassium iodide solution into the mixture; when half the iodide solution has been added, start the clock.

5. Record the time when the iodine/starch 'complex' colour appears, and *immediately* add a further 1·0 cm^3 portion of the sodium thiosulphate solution.

6. Repeat step *5* until six 1·0 cm^3 portions of the sodium thiosulphate solution have been added to obtain six times, all taken from the starting time.

After the addition of the fourth portion, start step *7*, since the fifth and sixth timings, without catalyst, take a comparatively long time (at least fifteen minutes each).

7. Repeat steps *1* to *6* five more times adding at stage *3*, from the appropriate burette, 0·5, 1·0, 2·5, 4·5 and 6·0 cm^3 respectively of the 0·001M copper(II) sulphate solution.

Observations

Enter the observations in a table headed as follows:

cm^3 of sodium thiosulphate solution added	time (in min) for the reappearance of iodine with the specified number of cm^3 of copper(II) sulphate solution					
	0	0·5	1·0	2·5	4·5	6·0

Calculations

1. Plot the volume of sodium thiosulphate solution added against the time for the reappearance of iodine, using no catalyst and successive different stipulated volumes of copper(II) sulphate solution, to obtain six curves with the same common axes.

2. Plot the time for the reappearance of iodine after the addition of the sixth 1·0 cm^3 portion of sodium thiosulphate solution against the volume of copper(II) sulphate solution employed.

3. Complete a table headed as follows:

cm^3 of copper(II) sulphate solution used	moles of copper(II) ions used	time (in min) when the sixth cm^3 of thiosulphate solution has been added	rate
	x	t_6	$1/t_6$

and plot the rate of the reaction against x.

Conclusion

Record all that can be deduced from the plots.

Questions

1. How many moles of copper(II) ions cause the rate of reaction to increase from 0·1 min^{-1} to 0·2 min^{-1}?

2. What is the concentration of copper(II) ions that causes this doubling of the rate of reaction?

9

EQUILIBRIA AND KINETICS

9.1 Equilibrium

A Sodium sulphate-water system

Requirements

50 g A.R. sodium sulphate crystals
60 cm^3 xylene
100 cm^3 beaker
-10°C to 110°C thermometer
test tube
boiling tube
stop-clock
500 cm^3 beaker, wire stirrer, boiling tube fitted with a two-holed rubber bung carrying a -5°C to 50°C (in 0·1°C) thermometer, a glass tube (50 cm × 0·25 cm) and a half metre rule held in position with two elastic bands, assembled as in the diagram on page 124

Procedure

1. Place two-thirds of a boiling tube of sodium sulphate crystals (c. 40 g) in the 100 cm^3 beaker, and wash well with half a test tube (c. 10 cm^3) of xylene.

2. Decant off the xylene and transfer the washed crystals to the second boiling tube.

3. Fill up this second boiling tube with xylene, ensuring that all air bubbles are removed, and carefully insert the rubber bung carrying the thermometer and glass tube.

4. Add water to nearly fill the 500 cm^3 beaker and heat until the temperature of the water is about 30°C.

5. Clamp the boiling tube and its attachments in the beaker of warm water.

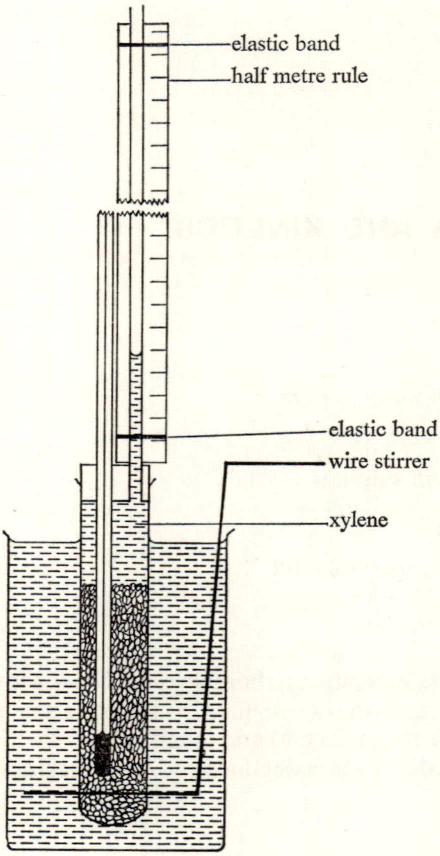

FIG. 22

6. While stirring constantly, heat the water *very slowly* (through about 1°C in ten minutes) and plot the height of xylene in the vertical tube against temperature for readings taken at one-minute time intervals, until the temperature reaches 34°C.

7. Stop heating, and in the manner described continue the plot until the temperature drops to 30° C.

Questions

1. How can the form of the graph be explained?

2. (*a*) What is the temperature of the equilibrium point?
(*b*) What is the usual name given to such an equilibrium point?

3. By what other means may this equilibrium temperature be determined?

4. (*a*) What is polymorphism?

(*b*) What is a dimorphic change and how might the heat energy associated with such a change be determined for the sodium sulphate-water system?

B *Determination of the equilibrium constant for the reaction between iron(II) and silver(I) ions*

Theory

For the reversible reaction

$$Fe^{2+}(aq)+Ag^{+}(aq) \rightleftharpoons Fe^{3+}(aq)+Ag(s)$$

the equilibrium constant is

$$K_c = \frac{[Fe^{3+}]}{[Fe^{2+}]\,[Ag^{+}]}$$

so
$$[Fe^{3+}] = K_c\,[Fe^{2+}]\,[Ag^{+}]$$

and a plot of $[Fe^{3+}]$ against $[Fe^{2+}]\,[Ag^{+}]$ gives a value of K_c from the slope.

Standard solutions of iron(II) and silver(I) ions in dilute sulphuric acid are used to prepare various reaction mixtures, and thus the initial concentrations of these ionic species are known; the mixtures are allowed to attain equilibrium. Titration of residual iron(II) ions (by standard potassium permanganate solution in the absence of metallic silver) leads to the concentrations of the other ions at equilibrium.

Requirements

Communal

400 cm³ 0·05M iron(II) sulphate in bench dilute sulphuric acid
400 cm³ 0·025M silver(I) sulphate in bench dilute sulphuric acid
450 cm³ 0·005M potassium permanganate solution
plastic bowl of water at the laboratory temperature or a thermostatic bath operating at 20°C

Individual

specified volumetric ware plus a 10 cm³ and a 50 cm³ pipette, 500 cm³ graduated flask and a 250 cm³ conical flask
cotton wool

Procedure

1. By reference to the table in the following section and by sensible choice of pipettes, make up a reaction mixture in a 250 cm³ conical flask. Stand the reaction flask in the constant temperature bath, plug the neck with cotton wool and leave overnight.

2. Attach the pipette filler to the 20 cm³ pipette.

3. For either A, B, C or D, withdraw a 20 cm³ portion of the equilibrium mixture, taking care not to disturb the precipitated silver, and titrate with the 0·005M potassium permanganate solution. Repeat to obtain concordant titres.

4. For either E, F or G,

(*a*) pipette 100 cm³ of the standard potassium permanganate solution into the 500 cm³ graduated flask and make up to the mark with purified water to obtain a solution which is 0·001M,

(*b*) repeat step *3* using the 0·001M potassium permanganate solution.

5. Class investigation leads to results for the other mixtures.

Observations and Calculations

1. Using class results, complete a table as shown (all concentrations in moles per litre):

	INITIALLY				AT EQUILIBRIUM				
	cm³ of 0·05M $FeSO_4$	cm³ of 0·025M Ag_2SO_4	$[Fe^{2+}]$	$[Ag^+]$	cm³ of 0·005M $KMnO_4$ for 20 cm³	$[Fe^{2+}]$	$[Fe^{3+}]$	$[Ag^+]$	$[Fe^{2+}][Ag^+]$
A	80	20							
B	70	30							
C	60	40							
D	50	50							
E	40	60							
F	30	70							
G	20	80							

2. Plot $[Fe^{3+}]$ against $[Fe^{2+}][Ag^+]$ and from it determine the value of K_c.

Questions

1. How is the expression for the equilibrium constant, K_c, obtained?

2. What are the units of K_c for the value determined?

3. Estimation of iron(II) ions in the equilibrium mixtures was carried out in the absence of silver. Why?
4. If the acid solution of silver(I) sulphate is not standard, how may its strength be determined?
5. For the equation

$$2Fe^{3+}(aq)+Zn(s) \rightleftharpoons 2Fe^{2+}(aq)+Zn^{2+}(aq)$$

what is the expression for K_c and how can this be determined experimentally?
6. Equilibrium constants K_c, are more accurately stated in terms of 'active masses' and not concentrations. Why?

9.2 Factors affecting rate of reaction

Questions

In this section, the reactant factors considered are particle size, concentration and temperature. What other factors affect the rates of chemical changes? How are rates affected by these additional factors?

A Variation of the particle size of a reactant

Requirements

2 g each of zinc foil, wool and powder
360 cm^3 30 per cent hydrochloric acid
Flowmaster pen or chinagraph pencil
100 cm^3 measuring cylinder
stop-clock
150 cm^3 bolt-head flask, delivery tube, a c. 700 cm^3 cordial bottle and a plastic bowl, assembled as shown:

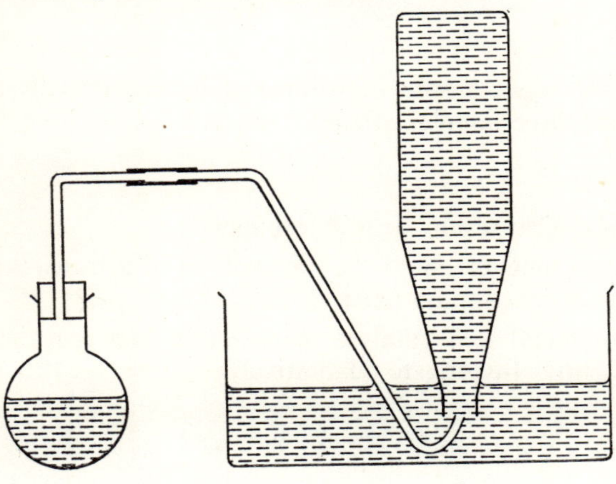

FIG. 23

Procedure

1. Graduate the cordial bottle by pouring, from the measuring cylinder, 100 cm^3 portions of water into it; after each addition, mark the water level with the special pen or pencil.

2. Assemble the apparatus as in figure 23.

3. Accurately weigh the piece of zinc foil and bend it into a V shape.

4. Pour two boiling tubes (c. 110 cm^3) of the hydrochloric acid into the reaction flask.

5. Simultaneously drop the zinc into the reaction flask and start the clock (i.e. zero time), and immediately insert the bung.

6. Record the times (from zero time) for the evolution of successive 100 cm^3 volumes of hydrogen liberated.

7. Wash out the reaction flask and refill the cordial bottle with water.

8. Repeat steps *4* to *7* employing exactly the same weight of metal with (*a*) zinc wool and (*b*) zinc powder.

Observations and Interpretation

1. Record the experimental data in a table with the following headings:

cm^3 of gas evolved	time in min for:		
	foil	wool	powder
100			
200			
300			

2. Plot, on the same axes, graphs of volume of hydrogen evolved against time for the three samples of zinc.

Questions

1. How can the graphical results be fully explained?

2. What other experiments could be devised to illustrate the dependence of rate of reaction on particle size?

3. What transition metal ions catalyse the reaction between zinc and hydrochloric acid? Investigate qualitatively.

*B Variation of the concentration of a reactant**

Theory

The time required for the sulphur to appear in the reaction

$$S_2O_3^{2-}(aq)+2H^+(aq) \longrightarrow SO_2(aq/g)+S(s)+H_2O(l)$$

is taken as a measure of the rate.

Requirements

75 cm³ 2·0M sodium thiosulphate solution
140 cm³ c. 1·0M nitric acid
two 10 cm³ pipettes
burette
glass stirring rod (at least 20 cm long)
three boiling tubes
three adhesive labels
stop-clock

Procedure

1. Label the burette 'purified water only', one 10 cm³ pipette 'thiosulphate only', and the other 10 cm³ pipette 'acid only'. Ensure that these pieces of apparatus are employed only in the manner indicated by their labels.

2. Pipette 10 cm³ of the 2·0M thiosulphate solution into a boiling tube.

3. Pipette 10 cm³ of the nitric acid into another boiling tube.

4. Simultaneously start the clock and add, stirring thoroughly, the acid to the thiosulphate solution.

5. Look down through the boiling tube, and when the first trace of milkiness appears stop the clock and record the time.

6. Thoroughly wash the glass rod and boiling tubes.

7. Repeat steps *2* to *6*.

8. Prepare a 1·5M thiosulphate solution by putting two pipetted 10 cm³ portions of the 2·0M thiosulphate solution into a boiling tube and adding, from the burette, 6·7 cm³ of purified water and mixing well.

9. For the 1·5M thiosulphate solution, repeat steps *2* to *7*.

10. Prepare a 1·0M thiosulphate solution by putting two pipetted 10 cm³ portions of the 2·0M thiosulphate solution into a boiling tube and adding, from the burette, 20 cm³ of purified water.

11. For the 1·0M thiosulphate solution, repeat steps *2* to *7*.

12. Prepare a 0·5M thiosulphate solution by adding 20 cm^3 of purified water, from the burette, to the remaining 20 cm^3 of the 1·0M sodium thiosulphate solution prepared in step *10*.

13. For the 0·5M thiosulphate solution, repeat steps *2* to *7*.

14. Repeat steps *12* and *13*, with the appropriate dilution, for 0·25M, 0·125M and 0·0625M solutions of sodium thiosulphate.

Observations and Calculations

1. Prepare a table of results with the following headings:

sodium thiosulphate solution		2·0M	1·5M	1·0M	0·5M	0·25M	0·125M	0·0625M
sec for first trace of milkiness	1st run							
	2nd run							
average time in sec	t							
rate	$1/t$							

2. Plot a graph of $1/t$ against concentration of the sodium thiosulphate solutions.

Questions

1. What conclusion can be drawn from the graph?

2. The points for the higher concentrations of sodium thiosulphate may deviate from the acceptable graph. Why is this?

C *Variation of the temperature of reactants**

Theory

The iodine initially liberated in the reaction

$$H_2O_2(aq)+2H^+(aq)+2I^-(aq) \longrightarrow I_2(aq)+2H_2O(l)$$

is removed by reaction with sodium thiosulphate:

$$2S_2O_3{}^{2-}(aq)+I_2(aq) \longrightarrow 2I^-(aq)+S_4O_6{}^{2-}(aq).$$

When all the thiosulphate has reacted, free iodine will appear, and the time taken for the appearance is a measure of the rate of reaction.

Requirements

550 cm³ acidified potassium iodide solution
30 cm³ c. 0·05M sodium thiosulphate solution
60 cm³ c. 1-volume hydrogen peroxide
starch indicator
250 cm³ beaker
100 cm³ pipette
10 cm³ pipette
burette
−10°C to 110°C thermometer
boiling tube
stop-clock

Procedure

1. Using the pipette, put 100 cm³ of the potassium iodide solution into the 250 cm³ beaker, and add five drops of starch indicator.

2. Using the burette, add exactly 5·0 cm³ of 0·05M sodium thiosulphate solution to the contents of the beaker.

3. Pipette a 10 cm³ portion of the hydrogen peroxide into the boiling tube.

4. Simultaneously (*a*) start the clock, (*b*) add the 10 cm³ portion of hydrogen peroxide to the contents of the beaker and stir with the thermometer, and (*c*) record the temperature of the solution.

5. As soon as the solution begins to darken and go blue, stop the clock and record the time.

6. Wash out the beaker with water.

7. Repeat steps *1* to *3* and then warm up the contents in the beaker to about 10°C above the initial temperature.

8. Repeat steps *4* to *6*.

9. Repeat steps *7* and *8* for temperatures about 20°C, 30°C and 40°C above the temperature recorded in *4* (*c*).

Observations and Calculations

1. Prepare a table with the following headings:

temperature			sec for appearance of I_2	rate	
°C	°A				
θ	T	$1/T$	t	$1/t$	$\log_{10} t$

2. Plot a graph of $1/t$ against θ.

3. Plot a graph of $\log_{10} t$ against $1/T$.

Conclusion

Comment on the form taken by the first graph.

Questions

1. What temperature rise will cause the rate of reaction to double?

2. Why does raising the temperature increase the rate of reaction?

3. Why does doubling the temperature more than double the rate of reaction?

4. Why does the acid not liberate sulphur from the thiosulphate as in experiment 9. 3B?

5. What are the parts played by iodide ions and iodine prior to the appearance of free iodine?

6. What is the significance of the plot of $\log_{10} t$ against $1/T$?

9.3 Investigations of rates and orders of reactions

Introduction

For an elementary reaction,	$A \longrightarrow$ product(s)
let moles per litre initially be	a
and moles per litre at time t be	a-x.

The rate, dx/dt, at which A decomposes is proportional to its concentration. Therefore,

$$\frac{dx}{dt} = k(a-x)$$

where k is the velocity (or rate) constant of the reaction. Integration gives

$$t = -\frac{1}{k}\log_e(a-x)+\text{constant}$$

when $t = 0$, $x = 0$ and $(\log_e a)/k$ is the constant. Thus

$$t = -\frac{1}{k}\log_e(a-x)+\frac{1}{k}\log_e a$$

$$= \frac{1}{k}\log_e\frac{a}{a-x}.$$

This equation may be expressed alternatively as

$$t = \frac{2{\cdot}303}{k} \log_{10} \frac{a}{a-x}$$

or exponentially as

$$(a-x) = ae^{-kt}.$$

If the *experimental observations* for a given reaction lead to a plot of t against $\log_{10}(a-x)$ being a straight line, then the reaction is said to be of the *first order*, and the slope of the graph will be $k/2{\cdot}303$.

When the reaction is half completed, the expression

$$t = \frac{1}{k} \log_e \frac{a}{a-x}$$

becomes

$$t_{0.5} = \frac{1}{k} \log_e \frac{a}{a-a/2}$$

$$= \frac{1}{k} \log_e 2.$$

That is, the time of half change, $t_{0.5}$, is independent of the initial concentration.

If this is apparent from the plot of $(a-x)$ against t, which should be one of exponential decay, then the reaction is confirmed as being of the first order.

For a reaction

$$\mathrm{A+B} \longrightarrow \text{product(s)}$$

where the initial concentration, a moles per litre, of each reactant is the same and the previously used notation is again employed,

$$\frac{\mathrm{d}x}{\mathrm{d}t} = k(a-x)^2$$

which leads to

$$t = \frac{1}{k} \cdot \frac{x}{a(a-x)}.$$

If a plot of t against $x/(a-x)$ is a straight line, then the reaction under *experimental observation* is of the *second order*, and the slope of the graph is $1/ka$. A time of half change which is inversely proportional to the initial concentration, i.e. $t_{0.5} = 1/ka$, confirms second order kinetics.

Questions

1. What is a pseudo first order reaction?

2. For a second order reaction (*a*) how is $dx/dt = k(a-x)^2$ converted to $t = x/\{ka(a-x)\}$, and (*b*) what is the expression for the time of quarter change?

3. What is meant by the molecularity of a chemical change?

A *Oxidation of formic acid by bromine**

Theory

In the reaction

$$Br_2(aq)+HCOOH(aq) \longrightarrow 2Br^-(aq)+2H^+(aq)+CO_2(g)$$

the concentration of formic acid (produced by the action of sulphuric acid on sodium formate) may be considered to be constant if the formic acid is present in a concentration which is large relative to that of bromine.

The variation in the concentration of bromine is determined by withdrawing successive equal volumes of the reaction mixture, treating with excess of potassium iodide solution (which stops the reaction), and titrating the liberated iodine with standard sodium thiosulphate solution.

Requirements

120 cm^3 bromine water
250 cm^3 sodium formate solution
120 cm^3 0·01M sodium thiosulphate solution
60 cm^3 0·001M sodium thiosulphate solution
120 cm^3 c. 0·5M sulphuric acid
200 cm^3 c. 0·1M potassium iodide solution
starch indicator
specified volumetric ware plus a 500 cm^3 conical flask
250 cm^3 measuring cylinder
stop-clock
plastic bowl

Procedure

1. Using the measuring cylinder, mix 200 cm^3 of the sodium formate solution and 100 cm^3 of the sulphuric acid in a 500 cm^3 flask.

2. Stand the flask in the plastic bowl of water at laboratory temperature (i.e. substitute for a constant temperature bath).

3. From the measuring cylinder, rapidly add, with swirling, 100 cm^3 of bromine water, recording the mean time of addition.

4. Attach the pipette filler to a 20 cm^3 pipette, and at five-minute intervals (after thirty minutes at ten-minute intervals) remove 20 cm^3 portions of the reaction mixture.

5. Add each portion to half a test tube (c. 10 cm^3) of potassium iodide solution, recording the mean time of addition.

6. Titrate the liberated iodine in each portion with the standard sodium thiosulphate solution (employ 0·01M until the titre is less than 3 cm^3, and thereafter 0·001M). Starch may be employed as indicator, in which case it should be added when the solution reaches a pale straw colour.

7. Proceed to withdraw samples until there is virtually no change in the titre.

8. Concurrent with steps *4* to *7*, plot titre against time.

Observations and Calculations

1. Tabulate the time/titration data as follows:

time in min	cm^3 of 0·01M sodium thiosulphate	
t	$(a-x)$	$\log_{10}(a-x)$

2. On the same axes, plot $(a-x)$ against t and $\log_{10}(a-x)$ against t, and comment on the relationships between the variables.

3. Deduce the value of the velocity constant and state its units.

B *Iodination of acetone in acid solution*

Requirements

60 cm^3 0·02M iodine in c. 0·2M potassium iodide solution
250 cm^3 0·01M sodium thiosulphate solution
100 cm^3 0·5M sodium bicarbonate solution
30 cm^3 1·0M sulphuric acid
30 cm^3 1·0M acetone in aqueous solution
starch indicator
specified volumetric ware plus two 10 cm^3 pipettes, a 25 cm^3 pipette, a 50 cm^3 pipette and two 250 cm^3 conical flasks
−10°C to 110°C thermometer
stop-clock
plastic bowl
five adhesive labels

Procedure

1. Label the conical flasks A, B and C respectively, and the 10 cm^3 pipettes X and Y.

2. Into flask A, pipette 50 cm^3 of the iodine/iodide solution, and clamp the flask in the bowl of water.

3. Into flask B, pipette 25 cm^3 of the acetone solution and 25 cm^3 of the sulphuric acid.

4. Using pipette X, put 10 cm^3 of the sodium bicarbonate solution into flask C.

5. Add all the contents of flask B to flask A and start the clock; record the time of half transfer. Shake the mixture thoroughly for about a minute.

6. Using the pipette labelled Y, immediately withdraw a 10 cm^3 sample of the reaction mixture, run it into flask C containing the sodium bicarbonate solution, and record the time for half transfer and the temperature of the reaction mixture.

7. Shake the mixture until no more carbon dioxide is evolved, and titrate the liberated iodine with the standard sodium thiosulphate solution. Starch may be employed as indicator when the solution reaches a pale straw colour.

8. Wash out flask C, and using pipette X transfer a 10 cm^3 portion of the sodium bicarbonate solution into it.

9. Repeat steps *6* to *8* at about seven and a half minute intervals until the reaction has been in progress for an hour or thereabouts.

10. After the last titration, immediately wash all the remaining reaction mixture down the drain.

Observations and Calculations

1. Complete a table with the following headings:

number of titration	time in min	tempera-ture in °C	cm^3 of 0·02M I_2 in 10 cm^3 of reaction mixture at zero time	cm^3 of 0·01M $Na_2S_2O_3$ used	cm^3 of 0·02M I_2 remain-ing in titration flask	cm^3 of 0·02M I_2 used in 10 cm^3 of reaction mixture
	t		a		$(a-x)$	x

2. On the same axes, plot graphs of $(a-x)$ against t and x against t.

Conclusion

Explain the forms taken by the graphs, and from the results state the order of the reaction with respect to iodine.

Questions

1. What is the equation for the rate-determining step in the iodination of acetone?

2. Why is a large excess of acetone employed?

3. In the preparation of the reaction mixture, why is the acetone/acid solution added to the iodine and not the other way round?

4. Why is the sodium bicarbonate solution used, and why is sodium carbonate solution an unsuitable alternative?

5. Why must the reaction mixture be washed away at the end of the reaction?

C *Titrimetric investigation of the catalytic decomposition of hydrogen peroxide**

Theory

The decomposition

$$H_2O_2(aq) \longrightarrow H_2O(l)+\tfrac{1}{2}O_2(g)$$

is slow at ordinary temperatures, but it is catalysed by colloidal manganese dioxide. The catalysis is achieved by adding a little potassium permanganate solution to an alkaline hydrogen peroxide solution (alkalinity is achieved and maintained by means of a borate buffer solution).

The course of the reaction is investigated by

(*a*) removing samples of the mixture at known time intervals,

(*b*) running them into dilute sulphuric acid and

(*c*) titrating the residual hydrogen peroxide with standard potassium permanganate solution.

Requirements

60 cm^3 borate buffer
250 cm^3 0·004M potassium permanganate solution
25 cm^3 c. 6-volume hydrogen peroxide
200 cm^3 bench dilute sulphuric acid
specified volumetric ware plus a 10 cm^3 pipette, a 50 cm^3 pipette and a 500 cm^3 conical flask
stop-clock
plastic bowl

Procedure

1. Pipette 10 cm^3 of the 6-volume hydrogen peroxide into the 500 cm^3 conical flask, and using the 50 cm^3 pipette add 150 cm^3 of purified water and 50 cm^3 of the borate buffer solution.

2. Clamp the flask in the plastic bowl containing a large volume of water at room temperature.

3. Pipette 10 cm^3 of the potassium permanganate solution into the 500 cm^3 conical flask; start the clock when half has been added (zero time) and mix well.

4. Put half a test tube (c. 10 cm^3) of bench dilute sulphuric acid in a 250 cm^3 conical flask. After the reaction mixture has stood for about five minutes, withdraw a 10 cm^3 pipetted portion and run it into the acid, recording the exact time when half has been added.

5. Immediately titrate the residual hydrogen peroxide in the withdrawn sample with the potassium permanganate solution.

6. Repeat steps *4* and *5* five more times at five minute intervals, and then at ten minute intervals until there is a negligible change in the titres.

7. During the ten minute intervals,

(*a*) pipette a 10 cm^3 portion of the original hydrogen peroxide into a 250 cm^3 graduated flask, make up to the mark with purified water and shake well;

(*b*) put a 10 cm^3 pipetted portion of this diluted hydrogen peroxide solution and half a test tube (c. 10 cm^3) of dilute sulphuric acid into a conical flask; titrate with the potassium permanganate solution and repeat to obtain concordant titres.

Observations and Calculations

1. During the course of the experiment, record all observations in a table with the following headings:

cm^3 of 0·004M permanganate equivalent to H_2O_2 initially in 10 cm^3 of reaction mixture, i.e. cm^3 from step 7 × 250/220	time in min	cm^3 of 0·004M permanganate equivalent to H_2O_2 left in 10 cm^3 of reaction mixture from step *5*	cm^3 of 0·004M permanganate equivalent to H_2O_2 decomposed from 10 cm^3 of reaction mixture
a	t	$(a-x)$	$a-(a-x) = x$

2. Plot the following graphs:

(*a*) $(a-x)$ against t,

(*b*) x against t; from this graph determine the slope of the curve, i.e. dx/dt, at several points (at least six) and tabulate thus:

$(a-x)$						
dx/dt						

(*c*) dx/dt against $(a-x)$.

Questions

1. What can be deduced from the graphs?

2. In what ways does the dilute sulphuric acid function in this investigation?

D *Gasometric investigation of the catalytic decomposition of hydrogen peroxide*

Theory

1. The decomposition

$$H_2O_2(aq) \longrightarrow H_2O(l) + \tfrac{1}{2}O_2(g)$$

is slow at room temperature, but is catalysed by the addition of tetramminecopper(II) ions. The reaction may be studied by measuring the volume of oxygen evolved at known intervals of time.

2. Hydrogen peroxide reacts slowly with iodide ions in acid solution:

$$H_2O_2(aq) + 2I^-(aq) + H^+(aq) \longrightarrow I_2(aq) + 2H_2O(l)$$

but this reaction is positively catalysed by molybdate ions. The liberated iodine may be titrated with standard sodium thiosulphate solution:

$$I_2(aq) + 2S_2O_3^{2-}(aq) \longrightarrow 2I^-(aq) + S_4O_6^{2-}(aq).$$

The maximum volume of oxygen which can be expected from the catalytic decomposition of the hydrogen peroxide can thus be calculated.

Requirements

18 cm^3 0·1M copper(II) sulphate solution
120 cm^3 0·1M sodium thiosulphate solution
2 cm^3 c. 3 per cent neutral ammonium molybdate solution
30 cm^3 c. 20-volume hydrogen peroxide
6 cm^3 c. 2M ammonium hydroxide solution
120 cm^3 c. M sulphuric acid
120 cm^3 c. 10 per cent potassium iodide solution
starch indicator
specified volumetric ware plus a 5 cm^3 pipette
25 cm^3 measuring cylinder with a glass stirring rod to fit
ignition tube
test tube
boiling tube
stop-clock
thread (c. 25 cm)
Ostwald gas burette with a 100 cm^3 graduated limb attached to a 250 cm^3 conical flask, as illustrated on page 141

Procedure

1. (*a*) Fill both limbs (up to their zero marks) of the Ostwald gas burette with water.

(*b*) Into the measuring cylinder pour 15 cm^3 of the copper(II) sulphate solution and 5 cm^3 of the ammonium hydroxide solution, and stir well.

(*c*) Wash out the 5 cm^3 pipette with purified water followed by a small volume of the approximately 20-volume hydrogen peroxide. Pipette 5 cm^3 of the hydrogen peroxide solution into the conical flask, add a boiling tube (c. 55 cm^3) of purified water and swirl well.

(*d*) Attach the length of thread to the ignition tube.

(*e*) Wash out the burette with purified water followed by a small volume of the complex solution. Use this burette to run 3 cm^3 of the complex solution into the clean ignition tube.

(*f*) Lower the ignition tube into the diluted hydrogen peroxide solution and secure it by inserting the bung in the neck of the conical flask (i.e. assembled as shown in the diagram).

(*g*) Using the Hofmann clip, allow water to run out of the gas burette until the water levels in both limbs are the same; read and record this level.

(*h*) Simultaneously tip the conical flask and swirl it well so that the decomposition commences, and start the clock.

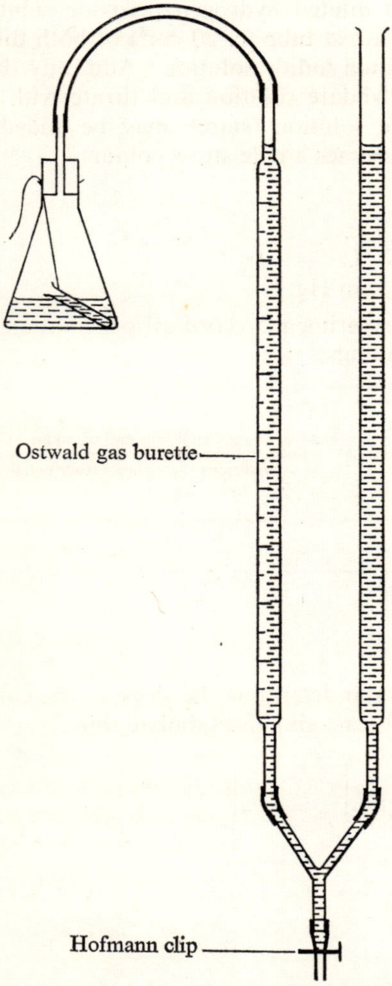

FIG. 24

(*i*) Just prior to the reaction having proceeded for an interval of two minutes, give the reaction flask two sharp shakes, and run out water from the open tube of the gas burette until at the two-minute time the water levels in the two limbs coincide. Immediately read and record this level.

(*j*) Repeat step (*i*) every two minutes until no more gas is evolved.

(*k*) Half an hour after the start of the reaction, ascertain the laboratory temperature and pressure.

2. (*a*) Pipette a 20 cm^3 portion of the hydrogen peroxide into a 250 cm^3 graduated flask, make up to the mark with purified water and shake well.

(*b*) Pipette 20 cm^3 of this diluted hydrogen peroxide solution into a conical flask, and add a test tube (c. 20 cm^3) of both dilute sulphuric acid and the potassium iodide solution. Add only three drops of the ammonium molybdate solution and titrate with the standard sodium thiosulphate solution (starch may be added as indicator when the solution assumes a pale straw colour).

Observations and Calculations

1. Laboratory temperature = °C
Barometric pressure = cm Hg

2. During the course of the experiment, record all observations in a table with the following headings:

titrimetrically estimated maximum cm^3, under laboratory conditions, of oxygen evolved from 5 cm^3 of approximately 20-volume H_2O_2	time in min	cm^3 of oxygen evolved under laboratory conditions	cm^3 of oxygen yet to be evolved under laboratory conditions
a	t	x	$(a-x)$

3. Plot the following graphs:

(*a*) $(a-x)$ against t,

(*b*) x against t; from this graph determine the slope of the curve, i.e. dx/dt, at several points (at least six) and tabulate thus:

$(a-x)$					
dx/dt					

(*c*) dx/dt against $(a-x)$.

Questions

1. What can be deduced from the graphs?

2. Why is it necessary to shake the reaction mixture before reading the volume of oxygen liberated?

3. What experimental observation may be made to illustrate the necessity of restricting the concentration of ammonium molybdate?

4. What other ammino complexes of transition metal ions are observed to catalyse this decomposition? Perform suitable test tube experiments.

E Acid hydrolysis of methyl formate*

Theory

The hydrolysis of methyl formate in aqueous solution,

$$\mathrm{H\,.\,COOCH_3}(aq/l)+\mathrm{H_2O}(l) \longrightarrow \mathrm{H\,.\,COOH}(aq)+\mathrm{CH_3OH}(aq)$$

and in the presence of a strong acid acting as a catalyst proceeds more rapidly than that of other simple esters.

The variation in the concentration of acid present is followed by the removal of successive small, equal volumes of the reaction mixture with subsequent titration, employing standard aqueous alkali.

Requirements

15 cm^3 methyl formate
220 cm^3 0·5M hydrochloric acid
250 cm^3 0·1M sodium hydroxide solution
phenolphthalein indicator solution
specified volumetric ware plus a 2 cm^3 pipette, a 10 cm^3 pipette, a 100 cm^3 pipette and a 500 cm^3 conical flask with bung to fit
stop-clock
plastic bowl

Procedure

1. Clamp the conical flask in the plastic bowl containing water at the laboratory temperature.

2. Using the 100 cm^3 pipette, put 200 cm^3 of the 0·5M hydrochloric acid in the 500 cm^3 conical flask.

3. Use the pipette filler and pipette 10 cm^3 of methyl formate into the acid, then start the clock.

4. Rapidly put the bung in the conical flask and shake well.

5. Immediately pipette 2 cm^3 of the reaction mixture into about 100 cm^3 of water in a 250 cm^3 conical flask and simultaneously record the time.

6. Immediately titrate with the 0·1M sodium hydroxide solution using phenolphthalein as indicator.

7. Repeat steps *5* and *6* at intervals of ten minutes until there is virtually no change in the titre.

Observations and Calculations

1. Tabulate the time/titration data in the following manner:

time in min	titre in cm^3 of 0·1M NaOH	final titre in cm^3 of 0·1M NaOH		
t	x	a	$(a-x)$	$\log_{10}(a-x)$

2. On the same axes, plot:

(*a*) x against t,
(*b*) $(a-x)$ against t,
(*c*) $\log_{10}(a-x)$ against t,

and comment on the relationships between the variables as indicated by the graphs.

3. Deduce the value of the velocity constant and state its units.

Questions

1. For the series of esters (*a*) $HCOOCH_3$, CH_3COOCH_3, $CH_3 . CH_2COOCH_3$, and (*b*) $HCOOCH_3$, $HCOOCH_2 . CH_3$, $HCOOCH_2 . CH_2 . CH_3$, how would the rate of acid hydrolysis through each series vary and why?

2. How could the rate of acid hydrolysis of methyl formate be used to determine the strengths of various acids?

F *Alkaline hydrolysis of methyl acetate**

Theory

The hydrolysis of methyl acetate in alkaline solution involves both ester molecules and hydroxyl ions:

$$CH_3COOC_2H_5(aq/l)+OH^-(aq) \longrightarrow CH_3COO^-(aq)+C_2H_5OH(aq).$$

The variation in the concentration of the hydroxyl ions can be followed by removal and estimation of successive small, equal volumes of the reaction mixture.

Requirements

1 cm^3 methyl acetate
140 cm^3 0·1M sodium hydroxide solution
300 cm^3 c. 0·025M hydrochloric acid
bromothymol blue indicator solution
specified volumetric ware plus a 5 cm^3 pipette, a 10 cm^3 pipette, a 100 cm^3 pipette, a 100 cm^3 graduated flask and a 500 cm^3 conical flask
−10°C to 110°C thermometer
boiling tube
stop-clock
plastic bowl

Procedure

1. (*a*) Fill the graduated flask with purified water to within no less than 2 cm of the graduation mark.

(*b*) Pipette 100 cm^3 of the sodium hydroxide solution into the 500 cm^3 conical flask and then wash out the pipette with purified water.

(*c*) Clamp both flasks in the plastic bowl containing water at the laboratory temperature.

2. Standardise the acid by titration of 5 cm^3 pipetted portions of standard alkali with it, employing bromothymol blue as indicator.

3. From a communal 2 cm^3 micro-burette, add exactly 0·79 cm^3 of methyl acetate to the water in the graduated flask and make up to the mark with purified water. Pour the contents of the graduated flask into the alkali contained in the 500 cm^3 conical flask, immediately start the clock, and swirl the mixture thoroughly.

4. (*a*) Pour two boiling tubes (c. 110 cm^3) of water into the 250 cm^3 conical flask.

(*b*) After the ester hydrolysis has proceeded for about three minutes, withdraw a 10 cm^3 pipetted portion of the reaction mixture. Run it into the water contained in the conical flask, and when half has been added record the time.

(*c*) At once, titrate the residual alkali in the diluted portion of the reaction mixture with the standardised acid, again employing bromothymol blue as indicator.

5. Repeat step *4* at intervals of five minutes until there is only slight variation in the volume of standardised acid required for consecutive withdrawals of the reaction mixture.

Observations and Calculations

1. Record the experimental data in the following manner (all concentrations in moles per litre); such a tabulation is only applicable to an initial reaction mixture of ester and alkali in equi-mole proportions:

Molarity of hydrochloric acid =

time in min	initial $[OH^-]$ and [ester]	cm^3 of HCl for 10 cm^3 of reaction mixture	$[OH^-]$ in reaction mixture			
t	a		$(a-x)$	$a-(a-x)$ $=x$	$\frac{x}{a-x}$	t
0	0·05		0·05 (i.e. $=a$)			

2. Plot $x/(a-x)$ against t, and from the slope of the graph obtain a numerical value of the velocity constant, k.

Questions

1. Why is each portion of the reaction mixture run into about 110 cm^3 of water?

2. What are the units of k, the velocity constant?

3. What is the time of half change for this hydrolysis reaction?

4. If unequal mole proportions of ester and alkali are used, what expression may be derived for k (calculate from first principles)?

5. The relative rates for the alkaline hydrolysis of a series of acetate esters, $CH_3 . COOR$, in aqueous solution at 25°C are

R	CH_3	$CH_3 . CH_2$	$(CH_3)_2 . CH$	$(CH_3)_3 . C$
relative rate	1	0·601	0·146	0·0084

Whereas for the alkaline hydrolysis of a series of ethyl esters, $R^1 . COOC_2H_5$, in 88 per cent aqueous ethanol at 30°C, the relative rates are

R^1	CH_3	$CH_3 . CH_2$	$(CH_3)_2 . CH$	$(CH_3)_3 . C$
relative rate	1	0·470	0·100	0·0105

What explanations can be given for these relative rates?

G *Oxidation of oxalic acid by permanganate ions*

Requirements

60 cm^3 0·012M manganese(II) sulphate solution
240 cm^3 c. 0·004M potassium permanganate solution
120 cm^3 c. 0·4M oxalic acid solution
250 cm^3 c. 0·01M sodium thiosulphate solution
240 cm^3 c. 0·005M sulphuric acid
260 cm^3 c. 1 per cent potassium iodide solution in c. 0·5M sulphuric acid
starch indicator
specified volumetric ware plus a 50 cm^3 pipette, a 100 cm^3 pipette and a 500 cm^3 conical flask
test tube
stop-clock
plastic bowl

Procedure

1. Into the 500 cm^3 conical flask, clamped in a plastic bowl of water at the laboratory temperature, pipette 100 cm^3 of the potassium permanganate solution, 100 cm^3 of the sulphuric acid and 50 cm^3 of water.

2. Put half a test tube (c. 10 cm^3) of the acidified potassium iodide solution into the 250 cm^3 conical flask.

3. Pipette 50 cm^3 of the oxalic acid solution into the contents of the 500 cm^3 conical flask; start the clock and record the time of half addition.

4. Immediately swirl the mixture in the 500 cm^3 conical flask and pipette 20 cm^3 of it into the 250 cm^3 conical flask containing the potassium iodide solution; again record the time of half addition.

5. Titrate the liberated iodine with the sodium thiosulphate solution. Starch may be used as indicator when the iodine solution is a pale straw colour.

6. Wash out the 250 cm^3 conical flask.

7. Repeat steps *2* and *4* to *6* at five-minute intervals for one hour.

8. Repeat the whole experiment, but this time (*a*) make up the original mixture in step *1* with the following pipetted portions: 100 cm^3 of the potassium permanganate solution, 100 cm^3 of sulphuric acid and 50 cm^3 of manganese(II) sulphate solution; and (*b*) carry out a second run by repeating steps *2* to *7* with this new mixture.

Observations and Calculations

1. Construct the following table, and enter the time and volume observations:

FIRST RUN			SECOND RUN		
time in min	cm³ of sodium thiosulphate	rate	time in min	cm³ of sodium thiosulphate	rate
t	v	dv/dt	t^1	v^1	dv^1/dt^1

2. For the first run,

(*a*) plot v against t,

(*b*) by construction and measurement, determine the slopes, dv/dt, at the times t, and enter these in the table,

(*c*) plot dv/dt against t.

3. Repeat step *2* for the second run.

Questions

1. How can the graphs be explained?

2. What is the ionic equation for the reaction between permanganate ions and iodide ions in acid conditions?

H Decomposition of oxalic acid by concentrated sulphuric acid

Requirements

Communal

12 g oxalic acid crystals (fine)
1000 cm³ concentrated sulphuric acid
50 cm³ pipette and pipette filler
2000 cm³ thick, hard glass, round-bottomed flask with water bath to fit
1000 cm³ measuring cylinder
100 cm³ beaker
stop-clock

Individual

250 cm³ c. 0·02M potassium permanganate solution
specified volumetric ware
two boiling tubes

Procedure

1. The communal reaction mixture is prepared by (*a*) weighing about 12 g of the oxalic acid into the 100 cm^3 beaker and transferring it to the round-bottomed flask, (*b*) adding, from the measuring cylinder, about 1000 cm^3 concentrated sulphuric acid and (*c*) heating the mixture on the water bath until the oxalic acid crystals seem to have disappeared.

2. During the ensuing two hours, a total of at least twenty samples should be successively and speedily removed from the bubbling reaction mixture: this may be achieved by using the 50 cm^3 pipette and pipette filler.

3. Each individual carries out the following instructions:

(*a*) Put two boiling tubes (c. 110 cm^3) of water into the 250 cm^3 conical flask and weigh.

(*b*) Employ the 50 cm^3 pipette and pipette filler to transfer about 20 cm^3 of the bubbling reaction mixture into the water already in the weighed conical flask (this is best achieved by dipping the end of the pipette into the water to avoid spluttering on running out the hot reaction mixture). When about half the volume has been added, note the time (the time for the first sample provides the zero time for the whole experiment).

After using the pipette arrangement, stand it in the washed-out measuring cylinder to avoid acid drips.

(*c*) Reweigh the conical flask and contents.

(*d*) Titrate the oxalic acid in the sample with the potassium permanganate solution under normal oxalate conditions.

(*e*) Wash out the conical flask and dry the outside of it.

(*f*) Record all results in the communal table of observations.

(*g*) As soon as it is convenient, repeat steps *3* (*a*) to (*f*) and continue to do so within the given time limit of two hours.

Observations and Calculations

1. Tabulate the time/titration data as follows:

number of sample	time in min	g of sample	cm^3 of $KMnO_4$ required	cm^3 of $KMnO_4$ for 10 g sample	
	t			$(a-x)$	$\log_{10}(a-x)$

2. On the same axes, plot $(a-x)$ against t and $\log_{10}(a-x)$ against t, and comment on the relationships between the variables as indicated by the graphs.

3. Deduce the value of the velocity constant and state its units.

Questions

1. (*a*) From the titration volume at zero time, what time intervals have elapsed for this titration volume to drop to one-half, one-quarter and one-eighth of the volume at zero time?

(*b*) What does this tell about the nature of the reaction investigated?

2. Why does the rate of bubbling in the reaction mixture increase when it is sucked up into the pipette?

3. What are the equations for the effect of concentrated sulphur acid on oxalic acid?

9.4 Determination of energies of activation

Introduction

The Arrhenius equation shows that the velocity (or rate) constant, k, of a reaction varies inversely with the absolute temperature, T:

$$\log_e k = -\frac{E}{RT} + \text{constant}$$

where E is the energy of activation of the reaction and R is the perfect gas constant.

The time, t, taken to complete a fraction of a reaction is inversely proportional to k:

$$t = \frac{1}{k}.$$

Hence

$$\log_e t = -\log_e k + \text{constant}'.$$

Thus

$$-\log_e t = -\frac{E}{RT} + \text{constant}''$$

$$\log_{10} t = \frac{E}{2{\cdot}303\ RT} + \text{constant}'''.$$

If a reaction is carried out at different temperatures, then the plot of $\log_{10} t$ against $1/T$ should give a straight line (thus verifying the Arrhenius equation) of slope $E/2{\cdot}303\ R$ (hence to the value of the energy of activation).

A Reaction between bromide and bromate ions in acidified aqueous solution

Theory

The ionic equation for the reaction is

$$BrO_3^-(aq) + 5Br^-(aq) + 6H^+(aq) \longrightarrow 3Br_2(aq) + 3H_2O(l).$$

The time taken for a given fraction of the reaction to take place is found by adding a slight stoichiometric deficiency of phenol and a little methyl red indicator. Bromine reacts rapidly with phenol:

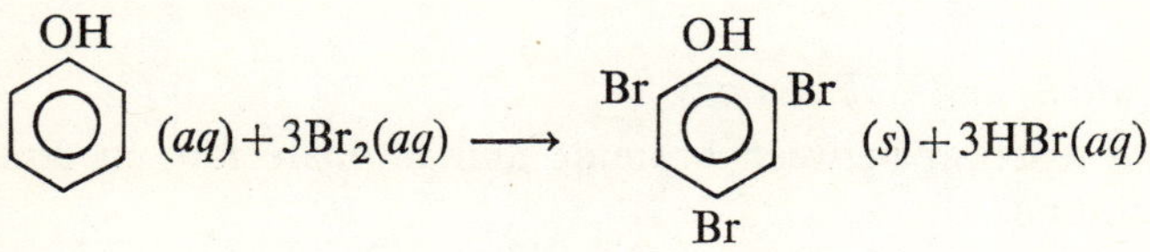

and as soon as this bromination is complete, any further bromine formed, by the interaction of bromide and bromate ions, destroys the indicator colour.

Requirements

60 cm^3 bromide/bromate solution (0·08M potassium bromide and 0·016M potassium bromate)
60 cm^3 0·0125M aqueous phenol solution
150 cm^3 c. 0·5M sulphuric acid
methyl red indicator solution
specified volumetric ware plus a 10 cm^3 pipette, a 25 cm^3 pipette and two 250 cm^3 conical flasks†
−10°C to 110°C thermometer
stop-clock
thermostatic bath†

Procedure

1. Adjust the controls of the thermostatic bath to maintain a bath temperature of 20°C.

2. (*a*) Put the aqueous phenol solution into the burette and then run 10 cm^3 of it into a 250 cm^3 conical flask.

(*b*) Add a 10 cm^3 pipetted portion of the bromide/bromate solution and four drops of methyl red solution.

3. Pipette 25 cm^3 of the sulphuric acid into the second conical flask.

4. Stand both conical flasks in the thermostatic bath, and allow the flasks and their contents to attain the set temperature.

5. Simultaneously start the clock and rapidly pour the acid into the bromide/bromate-phenol solution.

† If a thermostatic bath is not available, a litre beaker full of water may be either heated or cooled, by the addition of pieces of ice, to the required temperature, in which case boiling tubes should be employed instead of conical flasks.

6. Agitate the mixture and record the time, t sec, for the complete loss of the indicator colour.

7. Repeat steps *1* to *6* for each of the following temperatures: 30°C, 40°C, 50°C and 60°C.

Observations and Calculations

For the reaction between bromide and bromate ions in aqueous acid solution,

(*a*) complete a table with the following headings:

temperature of bath			time in sec	
°C	°A			
	T	$1/T$	t	$\log_{10} t$

(*b*) plot $\log_{10} t$ against $1/T$,

(*c*) (*i*) from the slope of the curve, obtain the value of $\dfrac{E}{2{\cdot}303\,R}$ (where R = 1·987 cal/mole deg),

(*ii*) evaluate E, the energy of activation, in kilocalories per mole.

Questions

1. For the bromide/bromate solution,

(*a*) what is the significance of the stated molarities of the constituents?

(*b*) what is the potential molarity with respect to bromine?

2. (*a*) Why is a slight stoichiometric deficiency of phenol employed?

(*b*) What is the systematic name of the bromination product of phenol?

B *Reaction between persulphate and iodide ions in aqueous solution*

Theory

The ionic equation for the reaction between persulphate ions and a large excess of iodide ions in aqueous solution is

$$S_2O_8^{2-}(aq)+2I^-(aq) \longrightarrow 2SO_4^{2-}(aq)+I_2(aq).$$

If the reaction is carried out in the presence of a small concentration of thiosulphate ions, then as soon as all the thiosulphate ions have been oxidised:

$$2S_2O_3^{2-}(aq)+I_2(aq) \longrightarrow S_4O_6^{2-}(aq)+2I^-(aq)$$

the 'free' iodine will produce the typical blue colour with starch.

Requirements

150 cm³ 0·01M potassium persulphate solution
70 cm³ 0·01M sodium thiosulphate solution
150 cm³ c. 0·5M potassium iodide solution
starch indicator
ice
10 cm³ pipette
two 20 cm³ pipettes
two 250 cm³ conical flasks†
−10°C to 110°C thermometer
stop-clock
thermostatic bath†

Procedure

1. Adjust the controls of the bath to maintain a temperature of 20°C.

2. Pipette 20 cm³ of the potassium iodide solution and 10 cm³ of the sodium thiosulphate solution into a conical flask.

3. Using the second 20 cm³ pipette, put 20 cm³ of the potassium persulphate solution and a few drops of starch indicator into the second conical flask.

4. Stand both conical flasks in the thermostatic bath and allow the flasks and their contents to attain the set temperature.

5. Simultaneously start the clock and rapidly pour the iodide/thiosulphate solution into the persulphate/starch solution.

6. Agitate the mixture and record the time, t sec, at which the blue colour of the iodine-starch appears.

7. Repeat steps *1* to *6* for each of the following temperatures: 10°C, 15°C, 25°C, 30°C and 35°C (when the temperature of the thermostatic bath has to be lowered, merely add pieces of ice).

Observations and Calculations

Repeat steps (*a*) to (*c*) of the corresponding section of experiment 9.4A to obtain the value of the energy of activation for the reaction between persulphate and iodide ions in aqueous solution.

† See footnote on page 151.

10

DOUBLE SALTS

Introduction

On crystallisation of an aqueous solution of two salts, it is usual for the salts to crystallise separately and in the order of increasing solubility. Both salts, however, may crystallise in one lattice as a double salt which only exists as such in the solid state. The composition of the crystals of a double salt is in accordance with the Law of Constant Composition. An aqueous solution of a double salt acts as a solution of the two constituent salts in the appropriate proportions.

Double salts are of three types. They may contain either, commonly, two different cations (apart from H^+), e.g. carnallite, $KCl \cdot MgCl_2 \cdot 6H_2O$; or less commonly, two different anions (apart from OH^- and O^{2-}), e.g. matlockite, $PbCl_2 \cdot PbF_2$; or rarely, two different cations and two different anions, e.g. kainite, $KCl \cdot MgSO_4 \cdot 3H_2O$.

10.1 Preparation of ammonium iron(II) sulphate crystals, $(NH_4)_2SO_4 \cdot FeSO_4 \cdot 6H_2O$

Requirements

6·0 g iron(II) sulphate crystals (fine)
3·0 g ammonium sulphate crystals (fine)
3 cm^3 bench dilute sulphuric acid
100 cm^3 beaker
25 cm^3 evaporating basin
7·5 cm diameter watch glass
ignition tube
test tube
filter funnel
glass stirring rod
three pieces of white blotting paper (c. 8 cm × 8 cm)
tile

Procedure

1. Weigh accurately about 5·0 g of the iron(II) sulphate and 2·4 g of the ammonium sulphate into the 100 cm^3 beaker. Add half an ignition tube (c. 2 cm^3) of bench dilute sulphuric acid and one-third of a test tube (c. 7 cm^3) of purified water.

2. Gently heat the contents of the beaker and stir until the crystals have dissolved.

3. Filter the solution through a previously moistened filter paper into the evaporating basin; cover with the watch glass and leave the solution to crystallise.

4. When crystallisation is complete, pour off the mother liquor and wash the crystals twice, using a half ignition tube portion (c. 2 cm^3) of purified water.

5. Press the crystals between two pieces of blotting paper and then transfer them to the third piece of blotting paper; set aside to dry on the tile and cover with the inverted dry beaker.

6. Weigh the dry crystalline product.

Questions

1. What is the percentage yield of ammonium iron(II) sulphate crystals?

2. Why is dilute sulphuric acid used in the preparation?

3. How may the percentage purity of the product be determined titrimetrically?

0.2 Ammonium copper(II) sulphate, $(NH_4)_2SO_4 . CuSO_4 . 6H_2O$

A *Preparation of crystals*

Requirements

15 g copper(II) sulphate crystals (fine)
7·5 g ammonium sulphate crystals (fine)
two 100 cm^3 beakers
250 cm^3 beaker
test tube
Buchner funnel and flask with attachments for suction filtration
glass stirring rod
piece of white blotting paper (c. 8 cm × 8 cm)
tile

Procedure

1. Accurately weigh 12·45 g of the copper(II) sulphate crystals into one 100 cm^3 beaker, add a test tube (c. 20 cm^3) of purified water, and warm and stir to obtain a solution.

2. Repeat step *1* employing exactly 6·6 g of the ammonium sulphate instead of the copper(II) sulphate crystals.

3. Pour the hot ammonium sulphate solution into the hot copper(II) sulphate solution. With the stirring rod, alternately stir and rub the inside of the beaker until crystals appear, and then set aside to crystallise further.

4. Filter off the crystals using the Buchner funnel and wash them twice with two quarter test tube portions (c. 5 cm^3) of purified water. Suck the crystals as dry as possible and then transfer them to the piece of blotting paper; allow to air dry between the tile and the inverted 250 cm^3 beaker.

5. Weigh the dry crystals.

Questions

1. What is the percentage yield of the product?

2. What is the significance of the chosen weights of copper(II) sulphate and ammonium sulphate crystals?

3. Why is tetramminecopper(II) sulphate (cuprammonium sulphate) not formed in this preparation?

B Estimation of percentage purity

Theory

When an excess of iodide ions is added to a solution containing copper(II) ions, copper(I) iodide is precipitated and iodine is liberated quantitatively:

$$2Cu^{2+}(aq)+4I^-(aq) \longrightarrow 2CuI(s)+I_2(aq).$$

The precipitate of copper(I) iodide does not interfere with the subsequent estimation of the iodine by titration with standard sodium thiosulphate solution in very slightly acid solution:

$$I_2(aq)+2S_2O_3^{2-}(aq) \longrightarrow S_4O_6^{2-}(aq)+2I^-(aq).$$

Requirements

12 g ammonium copper(II) sulphate crystals
100 cm^3 0·1M sodium thiosulphate solution
100 cm^3 c. 10 per cent potassium iodide solution
10 cm^3 bench dilute sodium carbonate solution
10 cm^3 bench dilute acetic acid
starch indicator
specified volumetric ware
ignition tube
test tube

Procedure

1. (*a*) Weigh accurately, into a 100 cm^3 beaker, about 10·0 g of the ammonium copper(II) sulphate and dissolve in a little purified water. Transfer the solution and washings to a 250 cm^3 graduated flask.

(*b*) Add dilute sodium carbonate solution to form a slight permanent precipitate which persists on shaking. Dissolve this precipitate by adding the minimum volume of dilute acetic acid.

(*c*) Make up the solution to the mark with purified water and shake well.

2. Pipette 20 cm^3 of the solution into a conical flask and add a test tube (c. 20 cm^3) of the potassium iodide solution. From a burette, run in the 0·1M sodium thiosulphate solution until the colour of the liberated iodine just disappears.

3. Repeat step *2*, adding half an ignition tube (c. 2 cm^3) of starch solution just prior to the end point, to obtain concordant titres.

Calculations

1. Determine the percentage of copper in (*a*) the prepared sample and (*b*) a theoretically pure specimen of ammonium copper(II) sulphate hexahydrate.

2. Calculate the percentage purity of the prepared double salt.

Questions

1. What is the name and formula of the precipitate obtained on adding sodium carbonate solution to the solution of copper(II) ions?

2. Why is the starch indicator only added just prior to the end point?

10.3 Preparation of aluminium potassium sulphate crystals, $Al_2(SO_4)_3 . K_2SO_4 . 24H_2O$

Requirements

1·2 g aluminium foil
4·0 g potassium sulphate crystals (fine)
2·5 cm^3 c. 0·1M solution of mercury(II) sulphate in bench dilute sulphuric acid
5 cm^3 concentrated sulphuric acid
250 cm^3 beaker
75 cm^3 evaporating basin
9 cm diameter watch glass
ignition tube
boiling tube
filter funnel
glass stirring rod
piece of glazed paper (c. 8 cm × 8 cm)
three pieces of white blotting paper (c. 8 cm × 8 cm)
tile

Procedure

1. Weigh accurately about 1·0 g of the aluminium foil.

2. Pour three-quarters of a boiling tube (c. 40 cm^3) of purified water into the beaker and then add, with stirring, an ignition tube (c. 4 cm^3) of concentrated sulphuric acid.

3. Add the aluminium foil and half an ignition tube (c. 2 cm^3) of the mercury(II) sulphate solution. Warm the mixture gently to start the reaction (further heating should be unnecessary).

4. When all the aluminium has dissolved, filter the solution into the evaporating basin.

5. Crease the piece of glazed paper lengthways, and weigh accurately on it about 3·4 g of potassium sulphate. Add these crystals to the solution in the evaporating basin and warm to dissolve.

6. Cover the solution with the watch glass and leave to crystallise.

7. When crystallisation is complete, filter off the mother liquor and wash the crystals with two separate ignition tube portions (c. 4 cm^3) of purified water.

8. Press the filter paper and crystals between two pieces of blotting paper. Transfer the crystals to the dry piece of blotting paper, and set aside to dry between the tile and the inverted dry beaker.

9. Weigh the dry crystals.

Questions

1. What is the percentage yield of potash alum?
2. Why is mercury(II) sulphate added? Could mercury(II) chloride be used instead?
3. What are the ionic equations of the reactions which have taken place during the preparation?
4. For potash alum, how may the following gravimetric estimations be carried out in practice: (*a*) aluminium, (*b*) potassium and (*c*) sulphate contents?
5. How may the aluminium content of potash alum be estimated titrimetrically?
6. How may potash alum be prepared starting with the reaction between aluminium and potassium hydroxide solution?

10.4 Chromium(III) potassium sulphate, $Cr_2(SO_4)_3 . K_2SO_4 . 24H_2O$

A Preparation of crystals

Requirements

12 g potassium dichromate crystals (fine)
25 cm^3 industrial alcohol
20 cm^3 concentrated sulphuric acid
250 cm^3 conical flask
100 cm^3 evaporating basin
9·5 cm diameter watch glass
10 cm^3 measuring cylinder
test tube
boiling tube
$-10°C$ to $110°C$ thermometer
teat pipette
plastic bowl
three pieces of white blotting paper (c. 8 cm × 8 cm)
tile

Procedure

1. Weigh accurately about 10·0 g of the potassium dichromate into the conical flask, add two-thirds of a boiling tube (c. 40 cm^3) of purified water and warm to dissolve.
2. Stand the conical flask in the plastic bowl of cold water and add cautiously, while stirring with the thermometer, 14 cm^3 of concentrated sulphuric acid from the measuring cylinder.
3. Cool this reaction mixture in a stream of cold water to 30°C.
4. While again stirring, add, using the teat pipette, a test tube

(c. 20 cm^3) of the alcohol dropwise: do *not* let the temperature rise above 60°C during this addition.

5. Transfer the solution to the evaporating basin, cover with the watch glass and set aside to crystallise.

6. When crystallisation is complete, pour off the mother liquor and wash the crystals twice, using a third of a test tube (c. 7 cm^3) of purified water each time.

7. Press the crystals of chrome alum between two pieces of blotting paper and leave them to dry on the third piece, between the tile and inverted dry evaporating basin.

8. Weigh the dry product.

Questions

1. What is the percentage yield of chrome alum?

2. Why must the temperature, during the addition of alcohol, not rise above 60°C?

3. To what is the alcohol converted?

4. The crystals cannot be dried in a steam oven. Why?

6. What are the ionic equations for the reactions which have taken place during the preparation?

B Estimation of percentage purity

Theory

Chromium(III) ions in aqueous solution may be oxidised to dichromate by employing persulphate ions:

$$2Cr^{3+}(aq)+3S_2O_8{}^{2-}(aq)+7H_2O(l) \longrightarrow Cr_2O_7{}^{2-}(aq)+6SO_4{}^{2-}(aq)+14H^+(aq).$$

The dichromate ions formed may be estimated by adding excess standard iron(II) solution and determining the residual iron(II) ions with standard potassium dichromate solution.

Requirements

2·5 g chromium(III) potassium sulphate crystals
250 cm^3 0·02M potassium dichromate solution
350 cm^3 c. 0·1M ammonium iron(II) sulphate solution
25 cm^3 25 per cent ammonium persulphate solution
12 cm^3 bench silver(I) nitrate solution
50 cm^3 syrupy orthophosphoric acid†

† If N-phenylanthranilic acid is employed as indicator, the use of orthophosphoric acid is not necessary.

400 cm^3 bench dilute sulphuric acid
barium diphenylamine sulphonate indicator solution (See footnote on p. 160)
specified volumetric ware plus a 50 cm^3 pipette and a 250 cm^3 conical flask
ignition tube
test tube
boiling tube

Procedure

1. Weigh out accurately about 2·0 g of the chrome alum in a 100 cm^3 beaker and add a boiling tube (c. 55 cm^3) of purified water to dissolve the crystals.
2. Transfer the chrome alum solution and washings, employing no more than a test tube (c. 20 cm^3) of purified water, to the 250 cm^3 conical flask. Add a test tube (c. 20 cm^3) of the ammonium persulphate solution and half a test tube (c. 10 cm^3) of bench silver(I) nitrate solution.
3. Place the stem of a glass filter funnel in the mouth of the conical flask, gently boil the contents for about twenty minutes and then allow this prepared dichromate solution to cool.
4. Since step *3* will take at least half an hour, during this time,
(*a*) pipette 20 cm^3 of the iron(II) sulphate solution into the conical flask, and add a test tube (c. 20 cm^3) of bench dilute sulphuric acid and four drops of barium diphenylamine sulphonate solution,
(*b*) from a burette, run in the standard 0·02M potassium dichromate solution until the colour of the solution turns dark green. Then add two ignition tubes (c. 8 cm^3) of the orthophosphoric acid and proceed to the end point,
(*c*) repeat steps *4* (*a*) and (*b*) to obtain concordant results.
5. From step *3*, transfer the solution and washings, employing no more than a boiling tube (c. 55 cm^3) of purified water, to a 250 cm^3 graduated flask, make up to the mark with purified water and shake well to give solution X.
6. Pipette two 20 cm^3 portions of this solution into a conical flask, and add a 50 cm^3 pipetted portion of the standardised ammonium iron(II) sulphate solution, a boiling tube (c. 55 cm^3) of bench dilute sulphuric acid and four drops of barium diphenylamine sulphonate solution.
7. Run in the standard dichromate solution until the colour turns to dark green. Then add two ignition tubes (c. 8 cm^3) of the orthophosphoric acid and proceed to the end point as before.
8. Repeat steps *6* and *7* to obtain concordant results.

Calculations

1. For the following sequence, calculate the number of moles explicit in each interdependent statement:

(*a*) dichromate ions in the final titre,

(*b*) dichromate ions reduced by 50 cm^3 of the ammonium iron(II) sulphate solution,

(*c*) the difference between (*a*) and (*b*) which is equivalent to the dichromate ions in 40 cm^3 of solution X, hence to the dichromate ions in 250 cm^3 of solution X and the chromium(III) ions which produced them.

2. Assuming the sample of chrome alum weighed initially to be pure, calculate the number of moles of chromium(III) ions arising from it.

3. From the last two calculations, determine the percentage purity of the prepared chrome alum.

Questions

1. Why is silver(I) nitrate solution added in step *2*?

2. How is the excess persulphate removed?

3. Why is it necessary to add syrupy orthophosphoric acid when using barium diphenylamine sulphonate as indicator?

4. What is the ionic equation for the reaction between dichromate and iron(II) ions in acid solution?

10.5 Ammonium iron(III) sulphate, $(NH_4)_2SO_4 . Fe_2(SO_4)_3 . 24H_2O$

A Preparation of crystals

Requirements

12 g iron(II) sulphate crystals (fine)
3·0 g ammonium sulphate crystals (fine)
25 cm^3 bench dilute sulphuric acid
3·0 cm^3 concentrated nitric acid
100 cm^3 beaker
75 cm^3 evaporating basin
9 cm diameter watch glass
ignition tube
test tube
filter funnel

glass stirring rod
piece of glazed paper (c. 8 cm × 8 cm)
three pieces of white blotting paper (c. 8 cm × 8 cm)
tile

Procedure

1. Weigh accurately about 10·0 g of the iron(II) sulphate into the beaker, add a test tube (c. 20 cm^3) of bench dilute sulphuric acid and warm to dissolve the crystals.

2. Heat the solution until it is boiling and then slowly add half an ignition tube (c. 2 cm^3) of concentrated nitric acid. Gently boil the solution until the volume is reduced by one-third.

3. On the piece of glazed paper, creased lengthways, weigh accurately about 2·4 g of the ammonium sulphate. Add these crystals to the solution, stir to dissolve, and filter through a previously moistened filter paper into the evaporating basin.

4. Cover the solution with a watch glass and set it aside to crystallise.

5. When crystallisation is complete, pour off the mother liquor and wash the crystals with two separate half ignition tube portions (c. 2 cm^3) of purified water.

6. Press the crystals between two pieces of blotting paper and then transfer them to the remaining piece of blotting paper; leave to dry on the tile under the inverted dry evaporating basin.

7. Weigh the dry crystals.

Questions

1. What is the percentage yield of ferric alum?

2. What is the purpose of the gentle boiling in step *2*?

3. What are the ionic equations for the reactions which have taken place during the preparation?

B *Estimation of the percentage purity**

Theory

Iron(III) is slowly reduced by iodide ions:

$$2Fe^{3+}(aq) + 2I^-(aq) \rightleftharpoons 2Fe^{2+}(aq) + I_2(aq).$$

The equation is driven to the right by a large excess of iodide ions and is catalysed by copper(I) iodide. The iodine produced is estimated by standard sodium thiosulphate solution.

Requirements

12 g ammonium iron(III) sulphate
5 cm^3 copper(I) iodide suspension
100 cm^3 0·05M sodium thiosulphate solution
50 cm^3 10 per cent potassium iodide solution
60 cm^3 bench dilute sulphuric acid
specified volumetric ware
ignition tube
test tube
boiling tube
glass stirring rod

Procedure

1. Weigh accurately about 10 g of the ferric alum in a 100 cm^3 beaker. Add a boiling tube (c. 55 cm^3) of bench dilute sulphuric acid, warm, and stir to dissolve the crystals. Transfer the solution and washings to a 250 cm^3 graduated flask, and make up to the mark with purified water.

2. (*a*) Pipette 20 cm^3 of the iron(III) solution into the conical flask, and add a quarter of an ignition tube (c. 1 cm^3) of the copper(I) iodide suspension, half a test tube (c. 10 cm^3) of the potassium iodide solution and half a boiling tube (c. 30 cm^3) of purified water. Swirl well and allow to stand for ten minutes.

(*b*) Titrate slowly with the standard sodium thiosulphate solution, stand for a further five minutes before deciding upon the end point.

3. Repeat steps *2* (*a*) and (*b*) to obtain concordant titres.

Calculations

1. Determine the percentage of iron in (*a*) the prepared sample and (*b*) a theoretically pure specimen of ferric alum.

2. Calculate the percentage purity of the ferric alum.

Questions

1. Although the reaction mixture is allowed to stand for ten minutes before titration, why is a further five minutes allowed to elapse before the end point is decided?

2. By what other titrimetric methods is it possible to estimate iron(III) ions?

11

CO-ORDINATION COMPOUNDS

Introduction

There are a large number of co-ordination compounds, all of which may be considered to contain co-ordinate bonds. The co-ordinate bond or dative covalent bond is one in which a lone pair of electrons form the bond. An ion or molecule which supplies a lone pair of electrons for bond formation of this type is called a ligand.

Co-ordination compounds may be based on either a cation, an atom or a molecule. The first type is by far the most common for it includes all the complex ions, which may be either cationic, e.g. the diamminesilver(I) ion, $Ag(NH_3)_2^+$; or anionic, e.g. the hexacyanoferrate(II) or ferrocyanide ion, $Fe(CN)_6^{4-}$. Nickel tetracarbonyl, $Ni(CO)_4$, so important in the extraction of nickel, is one of a series of metal carbonyls which are co-ordination compounds centred on neutral metal atoms. An example of a molecule entering into co-ordination is provided by boron trifluoride, which unites with ammonia to form an ammine, $BF_3 . NH_3$.

In this chapter, complex ions have been named in accordance with the recommendation of the International Union of Pure and Applied Chemistry (I.U.P.A.C.).

Questions

1. What are the electronic structures of the co-ordination compounds referred to above?

2. What are the formulae of the metal carbonyls formed by the transition elements in the first long period of the Periodic Classification?

3. How may the ammine of aluminium chloride be formed, and what is its electronic structure?

11.1 Determination of the instability constant of the complex tri-iodide ion, I_3^-

Theory

For the equilibrium

$$KI_3(aq) \rightleftharpoons KI(aq) + I_2(aq)$$

i.e.

$$I_3^-(aq) \rightleftharpoons I^-(aq) + I_2(aq)$$

the concentrations of all three species must be ascertained in order to determine the value of the instability constant, k_i, for the complex tri-iodide ion,

$$k_i = \frac{[I^-]\,[I_2]}{[I_3^-]}.$$

This may be achieved by allowing mixtures of solutions of iodine in benzene and potassium iodide in water, under the normal laboratory temperature, to reach the following equilibrium:

$$\begin{array}{ll} I_2 & \text{benzene layer} \\ \updownarrow & \\ I_2 + I^- \rightleftharpoons I_3^- & \text{aqueous layer} \end{array}$$

in which the free iodine distributes itself between the two layers in accordance with the distribution coefficient

$$[I_2]_{\text{benzene}} / [I_2]_{\text{water}} = 130\dagger$$

Estimation of the concentration of iodine in the benzene layer leads to the concentration of free iodine in the aqueous layer, $[I_2]$. Estimation of the concentration of the total iodine in the aqueous layer, $[I_2]$ and $[I_3^-]$, followed by subtraction gives the concentration of the complex ion, $[I_3^-]$. The initial concentration of the potassium iodide is known; thus a further subtraction leads to the equilibrium concentration of the iodide ion, $[I^-]$.

Requirements

40 cm^3 0·1M sodium thiosulphate solution
100 cm^3 0·04M sodium thiosulphate solution
30 cm^3 solution of iodine in benzene (c. 40 g/1)
120 cm^3 aqueous solution of potassium iodide (c. 20 g/1)
25 cm^3 benzene (TOXIC: CARE)
starch indicator
specified volumetric ware plus a 50 cm^3 burette with glass tap
250 cm^3 separating funnel
250 cm^3 glass-stoppered reagent bottle

† As determined in experiment 2.2.

Procedure

1. Put the benzene, the solution of iodine in benzene, the aqueous potassium iodide solution, and water separately into four communal burettes (the first two should be glass-tapped). Use these burettes to prepare one of the following mixtures, in the separating funnel:

mixture	cm^3 benzene	cm^3 solution iodine in benzene	cm^3 potassium iodide solution	cm^3 water
A	19	6	50	50
B	12	13	67	33
C	6	19	84	16
D	0	25	100	0

2. Stopper the separating funnel, shake well and leave overnight (or longer, if possible).
3. Separate the layers in the mixture by running the aqueous layer into a conical flask and the benzene layer into the dry glass-tapped burette.
4. Estimate the total iodine in the aqueous layer by titration of two separate 40 cm^3 pipetted portions (2×20 cm^3 each) with the 0·04M sodium thiosulphate solution; when the colour of the iodine solution has paled markedly, starch may be employed as indicator.
5. Estimate the iodine in the benzene layer by titration of two separate 10 cm^3 buretted portions with the 0·1M sodium thiosulphate solution performed in the stoppered reagent bottle with constant shaking.
6. Class investigation leads to the results for the other mixtures.

Observations and Calculations

1. Complete a table, as shown, to cover all the mixtures, with concentration terms expressed in moles per litre:

mixture	cm^3 of 0·04M thiosulphate for 40 cm^3 of aqueous layer	cm^3 of 0·1M thiosulphate for 10 cm^3 of benzene layer	$[I_2]$	$[I_3^-]$	$[I^-]$	$[I_2][I^-]$	k_i

2. From the class results plot $[I_2][I^-]$ against $[I_3^-]$, and from the slope deduce the value of the instability constant, k_i.

11.2 Preparation of tetramminecopper(II) sulphate crystals, $[Cu(NH_3)_4]SO_4 . H_2O$†

Requirements

5·0 g copper(II) sulphate crystals
10 cm^3 0·88 ammonia
75 cm^3 industrial alcohol
25 cm^3 pipette
100 cm^3 beaker
9 cm diameter watch glass
test tube
glass stirring rod
Buchner flask and funnel with attachments for suction filtration
atmospheric desiccator containing self-indicating silica gel

Procedure

1. Weigh accurately about 4·0 g of the copper(II) sulphate into the beaker.

2. Add a test tube (c. 20 cm^3) of water, warm and stir the mixture to accomplish solution, and then cool by shaking the flask in a stream of water from a cold tap.

3. Cautiously add 0·88 ammonia with stirring until the precipitate formed initially has all dissolved.

4. From the pipette, add dropwise with stirring about 25 cm^3 of industrial alcohol.

5. Filter off the precipitate at the Buchner funnel and wash with two separate test tube portions (c. 20 cm^3 each) of industrial alcohol.

6. Transfer the product to the watch glass and dry for at least twenty-four hours in the atmospheric desiccator over silica gel.

7. Weigh the dry product.

Questions

1. What is the percentage yield for the preparation?

2. What are the ionic equations for (*a*) the precipitation reaction in step *3*, and (*b*) the complete preparation?

3. What is the purpose of the addition of industrial alcohol?

4. What different types of chemical bonding are present in the compound prepared?

† Hexamminenickel(II) sulphate may be obtained in a like manner.

1.3 Comparison of the reactions of copper(II) and tetramminecopper(II) ions in aqueous solution

Requirements

2·0 g copper(II) sulphate crystals
2·0 g tetramminecopper(II) sulphate
0·5 g disodium ethylenediaminetetracetic acid (i.e. E.D.T.A.)
normal laboratory reagents
semi-micro apparatus as employed for qualitative analysis

Procedure

1. Heat a match-head bulk of each solid in an ignition tube and observe.

2. Carry out the tests listed below on 1 cm^3 portions of each of the two copper solutions; make careful observations after the addition of each reagent.

(*a*) Pass hydrogen sulphide for at least one minute and then add dilute hydrochloric acid dropwise.

(*b*) Make dropwise addition of the following solutions:

(*i*) dilute sodium hydroxide,

(*ii*) potassium hexacyanoferrate(II) (i.e. potassium ferrocyanide) solution,

(*iii*) potassium cyanide solution (VERY POISONOUS : CARRY OUT IN A FUME CUPBOARD),

(*iv*) potassium iodide solution,

(*v*) potassium thiocyanate solution.

(*c*) Add solid E.D.T.A. until there is no further change in colour and centrifuge.

(*d*) Add about 3 cm of magnesium ribbon and allow to stand for five minutes before careful inspection.

Observations

Record all the information in two tables, one for the copper(II) ion solution and the other for the tetramminecopper(II) ion solution, with the following headings:

name of reagent	observations	name and formula of product containing copper

Conclusion

Comment on the stability of the tetramminecopper(II) complex ion.

11.4 Determination of the formula of the ammonia-copper(II) complex ion

Theory

When a good excess of ammonium hydroxide is added to an aqueous solution of copper(II) ions, the ammonia reacts to form a deep blue complex ion, $[Cu(NH_3)_x]^{2+}$, and the following equilibrium is set up:

$$Cu^{2+}(aq)+xNH_3(aq) \rightleftharpoons [Cu(NH_3)_x]^{2+}(aq).$$

The equilibrium may be examined by allowing the excess ammonia to distribute itself between the aqueous layer and chloroform, in accordance with the distribution coefficient:

$$[NH_3]_{water} \,/\, [NH_3]_{chloroform} = 25.$$

Estimation of the concentration of ammonia in the chloroform layer, and use of the distribution coefficient, leads to the concentration of free ammonia in the aqueous layer. Estimation of the total ammonia, free and complexed, in the aqueous layer gives the concentration of complexed ammonia. With a standard copper(II) ion solution, the number of ammonia molecules which have combined with one copper(II) ion can be deduced.

Requirements

30 cm^3 0·1M copper(II) sulphate solution
125 cm^3 0·5M hydrochloric acid
100 cm^3 0·025M hydrochloric acid
60 cm^3 c. 1·0M ammonium hydroxide
80 cm^3 chloroform
screened methyl orange indicator solution
specified volumetric ware plus a 50 cm^3 pipette
250 cm^3 separating funnel
100 cm^3 measuring cylinder
250 cm^3 stoppered reagent bottle

Procedure

1. Into the separating funnel, (*a*) pipette 20 cm^3 of the copper(II) sulphate solution, (*b*) pipette 50 cm^3 of the ammonium hydroxide to obtain a deep blue colour and finally (*c*) burette 70 cm^3 of chloroform.

2. Stopper the separating funnel and shake thoroughly for about three minutes, and then set the funnel aside for the two layers to separate.

3. Attach a pipette filler to a 20 cm^3 pipette.

4. Withdraw a 20 cm^3 portion of the aqueous layer and run it into a conical flask. Employ screened methyl orange as indicator and titrate with the 0·5M hydrochloric acid until the colour just turns purple.

5. Repeat step *4* with two further 20 cm^3 portions and then wash out the pipette.

6. With the pipette filler still attached, lower the tip of the pipette through the remaining aqueous layer and into the chloroform layer. Gently blow out any of the aqueous layer which has entered the pipette and transfer a 20 cm^3 portion of the chloroform layer into the reagent bottle. Estimate the ammonia present by titration with the 0·025M hydrochloric acid, employing screened methyl orange as indicator; stopper the bottle and shake thoroughly after each addition of acid.

7. Repeat step *6* with two further 20 cm^3 portions of the chloroform layer.

Calculations

Expressing all concentrations in moles per litre, calculate:

(*a*) the concentrations of (*i*) total ammonia in the aqueous layer, (*ii*) ammonia in the chloroform layer, (*iii*) free ammonia in the aqueous layer, (*iv*) complexed ammonia in the aqueous layer and (*v*) copper(II) ions in the aqueous layer;

(*b*) the ratio of the concentration of complexed ammonia to the concentration of copper(II) ions, and thus the formula of the complex ion.

Questions

1. What is the equation for the complete reaction between copper(II) ions and ammonium hydroxide in aqueous solution?

2. Why is a *good* excess of ammonia essential at the outset?

11.5 Preparation of bisglycinecopper(II), $Cu(C_2O_2NH_4)_2$

Requirements

18 g copper(II) carbonate
9·0 g glycine
25 cm^3 2:1 by volume industrial alcohol/water solution
25 cm^3 industrial alcohol
25 cm^3 acetone
two 100 cm^3 beakers
250 cm^3 beaker
500 cm^3 conical flask
350 cm^3 evaporating basin
15 cm diameter watch glass
test tube
1000 cm^3 can
two pieces of white blotting paper (c. 8 cm × 8 cm)
500 cm^3 Buchner flask and funnel (disc diameter 9 cm) with attachments for suction filtration

Procedure

1. Weigh 15·0 g of the copper(II) carbonate and 7·5 g of the glycine in separate 100 cm^3 beakers. Fill the 250 cm^3 beaker with purified water and employ this to transfer the two weighed samples to the conical flask.

2. Boil the mixture for at least fifteen minutes, and in the meantime heat the Buchner funnel by immersion in the can full of boiling water. Filter the hot liquid through the heated Buchner funnel into the Buchner flask.

3. Transfer the clear filtrate to the evaporating basin. Evaporate until crystals just begin to form at the edge of the solution, and then set aside to crystallise, covered with the watch glass.

4. Filter off the crystals using the Buchner funnel and wash successively with test tube portions (c. 20 cm^3) of (*a*) the alcohol/water solution, (*b*) alcohol and (*c*) acetone. Suck the crystals as dry as possible, and then press between the two pieces of blotting paper and allow to air dry.

Question

What is the structural formula of the product?

11.6 Determination of the formula of the ethylenediamine-tetracetate copper(II) complex ion*

Requirements

100 cm³ 0·04M copper(II) sulphate solution
100 cm³ 0·05M disodium ethylenediaminetetracetic acid (i.e. E.D.T.A.) solution
1 cm³ 1:1 by volume 0·88 ammonia/water solution
murexide indicator mixture
specified volumetric ware
test tube
teat pipette

Procedure

1. (*a*) Pipette 20 cm³ of the copper(II) sulphate solution into a conical flask and add a test tube (c. 20 cm³) of purified water.

(*b*) Using the teat pipette, add the ammonia solution dropwise until the precipitate which forms just redissolves.

(*c*) Swirl into the solution a pea-sized portion of murexide indicator and titrate with the E.D.T.A. solution.

2. Obtain concordant results for step *1*.

Questions

1. What is the ratio of the number of moles for the reaction between copper(II) ions and disodium ethylenediaminetetracetic acid?

2. What is the structural formula of the complex?

11.7 Verification of the formula of the bis(dimethylglyoximato)nickel(II) complex

Requirements

0·5 g A.R. nickel(II) sulphate crystals
125 cm³ 1 per cent alcoholic solution of dimethylglyoxime
5 cm³ concentrated hydrochloric acid
12 cm³ 1:5 by volume 0·88 ammonia/water solution
red litmus paper
100 cm³ beaker
1000 cm³ beaker
ignition tube

test tube
glass stirring rod
atmospheric desiccator containing self-indicating silica gel
Buchner flask and No. 4 sintered glass crucible with attachments for suction filtration
oven with thermostatic control

Procedure

1. Weigh accurately about 0·4 g of the nickel sulphate into the 100 cm^3 beaker. Transfer these crystals and washings to the 1000 cm^3 beaker, add an ignition tube (c. 4 cm^3) of the concentrated hydrochloric acid, and half fill the beaker with purified water.

2. With stirring, heat to almost boiling and, when solution is complete, add two boiling tubes (c. 110 cm^3) of the dimethylglyoxime solution.

3. (*a*) Again while stirring, add half a test tube (c. 10 cm^3) of the ammonia solution; this should produce a precipitate in an ammoniacal medium (spot test for this on the red litmus paper). Again heat to almost boiling for at least half an hour and then allow to cool.

(*b*) In the meantime, dry the clean crucible in the oven at 120°C, allow to cool in the desiccator, and weigh to constant weight.

4. Stir the cool mixture, filter off the product through the sintered glass crucible and wash the precipitate thoroughly with purified water.

5. Dry the precipitate and crucible in the oven at 120°C, cool in the desiccator and weigh to constant weight.

Questions

1. (*a*) What is the accepted equation for the reaction?

(*b*) What is the experimentally determined ratio of the number of moles of reactants for the precipitation of the dimethylglyoxime nickel(II) complex? Does this agree with the accepted ratio as quoted in the equation?

(*c*) What is the electronic structural formula of the precipitated complex?

2. (*a*) What are the characteristics, which may be observed on formation in a test tube, for the dimethylglyoxime complexes with (*i*) cobalt(II) ions and (*ii*) copper(II) ions?

(*b*) What is the constitutional difference between these complexes and the bis(dimethylglyoximato)nickel(II) complex?

11.8 Preparation of hexamminecobalt(III) chloride crystals, $[Co(NH_3)_6]Cl_3$

Reaction equation

$$2CoCl_2(aq)+2NH_4Cl(aq)+10NH_3(aq)+H_2O_2(aq) \longrightarrow 2[Co(NH_3)_6]Cl_3(s)+2H_2O(l)$$

Requirements

20 g cobalt(II) chloride crystals
15 g ammonium chloride
0·6 g activated charcoal
45 cm^3 0·88 ammonia
50 cm^3 20-volume hydrogen peroxide
30 cm^3 concentrated hydrochloric acid
70 cm^3 60 per cent industrial alcohol
70 cm^3 industrial alcohol
crushed ice
two 100 cm^3 beakers
250 cm^3 beaker
250 cm^3 conical flask
50 cm^3 burette
9 cm diameter watch glass
test tube
boiling tube
glass stirring rod
$-10°C$ to $110°C$ thermometer
plastic bowl
1000 cm^3 can
four pieces of white blotting paper (c. 8 cm × 8 cm)
Buchner flask and funnel with attachments for suction filtration
oven with thermostatic control

Procedure

1. Weigh accurately about 18·0 g of the cobalt(II) chloride crystals into one of the 100 cm^3 beakers.

2. Weigh about 12·0 g of the ammonium chloride into the other 100 cm^3 beaker and add half a boiling tube (c. 30 cm^3) of purified water. Stir and boil to obtain a solution and then dissolve the weighed portion of cobalt(II) chloride in this solution.

3. Put half an ignition tube of activated charcoal (c. 0·5 g) in the conical flask and pour the hot solution over it.

4. Wash out the beaker in which the mixed solution was made with two successive test tube portions (c. 20 cm^3 each) of 0·88 ammonia, pouring each in turn slowly into the conical flask.

5. Surround the conical flask with crushed ice contained in the plastic bowl. With the thermometer, stir the contents of the flask until the temperature falls below 10°C.

6. From the burette, while briskly agitating the flask in the ice, slowly add 45 cm^3 of the hydrogen peroxide to the reaction mixture, ensuring that the temperature remains below 10°C.

7. Put cold water in the can and stand the conical flask in it. Gradually heat the water in the bath to reach a temperature of about 60°C, and then maintain the temperature of the reaction mixture within the range 50°C to 60°C until the original dark red colour turns to a pale yellow brown with a pinkish tinge.

8. Again cool the conical flask and contents in the ice bath (until crystallisation is assumed to be complete).

9. Filter off the crude product at the Buchner funnel, and put it and the filter paper into about 150 cm^3 of purified water contained in the 250 cm^3 beaker. Add an ignition tube (c. 4 cm^3) of concentrated hydrochloric acid, and boil this mixture until the only solid remaining is charcoal and filter paper.

10. Filter the hot liquid into the Buchner flask.

11. Stand the conical flask in crushed ice, and add the filtrate and two test tubes (c. 40 cm^3) of concentrated hydrochloric acid. Stir and then leave to crystallise.

12. (*a*) Filter off the crystalline product at the Buchner funnel and retain the mother liquor; wash the crystals with a boiling tube (c. 55 cm^3) of the 60 per cent alcohol and then with the same volume of industrial alcohol. Suck the crystals as dry as possible, transfer them to the watch glass and complete the drying in the oven at 80°C.

(*b*) A second crop of crystals may be obtained by concentrating the mother liquor from step *12* (*a*) and repeating the procedure!

13. Weigh the dry product.

Questions

1. What is the ionic equation for the preparation?

2. What is the percentage yield for the preparation?

3. Why is activated charcoal added to the reaction mixture?

4. Cobalt(III) compounds are usually unstable: how may the stability of the product be explained?

11.9 Preparation of trisodium hexanitrocobaltate(III) crystals, $Na_3[Co(NO_2)_6]$

Requirements

3·5 g cobalt(II) nitrate crystals
8·0 g sodium nitrite
2·5 cm³ glacial acetic acid
25 cm³ industrial alcohol
12 cm³ acetone
100 cm³ beaker
7·5 cm diameter watch glass
ignition tube
test tube
glass stirring rod
−10°C to 110°C thermometer
two pieces of white blotting paper (c. 8 cm × 8 cm)
tile
filter tube (150 mm × 24 mm) with attachments so that it can be assembled in two ways, as in the diagrams:

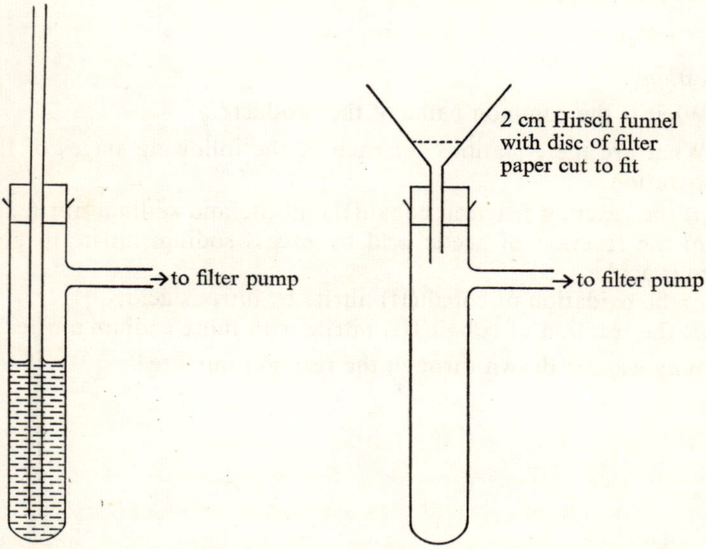

(*a*) FIG. 25 (*b*)

Procedure

1. Half fill the ignition tube (c. 2 cm^3) with glacial acetic acid, top up with purified water and, with a finger over the end, invert several times.

2. Weigh accurately into the beaker about 3·0 g of cobalt(II) nitrate and about 7·0 g of sodium nitrite. Add half a test tube (c. 10 cm^3) of purified water, stir with the thermometer, warm to 50°C to obtain a solution, and then add the diluted acetic acid.

3. Transfer the contents of the beaker to the test tube with the side arm and assemble as shown in figure 25 (*a*). Draw a current of air through the solution for at least twenty minutes.

4. Pour the solution back into the beaker and, while stirring, add half a test tube (c. 10 cm^3) of industrial alcohol.

5. Cover the beaker with the watch glass and allow to stand for at least an hour (the longer, the better—even overnight).

6. Assemble the apparatus as in figure 25 (*b*) employing very retentive filter paper. Filter off the crystalline product on the Hirsch funnel, and wash with half a test tube (c. 10 cm^3) of industrial alcohol and again with the same volume of acetone. Suck the crystals as dry as possible and then air dry between the pieces of blotting paper: place on the tile and cover with the watch glass.

Questions

1. What is the common name of the product?

2. What are the equations for each of the following stages of the preparation:

(*a*) the reaction between cobalt(II) nitrate and sodium nitrite,

(*b*) the reaction of acetic acid on excess sodium nitrite to give nitrous acid,

(*c*) the oxidation of cobalt(II) nitrite by nitrous acid,

(*d*) the reaction of cobalt(III) nitrite with more sodium nitrite?

3. Why was air drawn through the reaction mixture?

12

COVALENT COMPOUNDS

Introduction

Atomic spectra support the view that noble gas electronic configurations are exceptionally stable and that atoms, in chemical combination, tend to acquire these stable structures.

An alkali metal has one electron in excess of a noble gas structure and, as the result of energy changes in chemical reactions, may lose this outer electron. The halogen atoms need to gain one electron to achieve a similar state. In the reaction between sodium and fluorine, there is electron transfer between these two atoms of widely differing electronegativities. This results in the formation of positive sodium ions and negative fluoride ions which are held together non-directionally by electrostatic attraction:

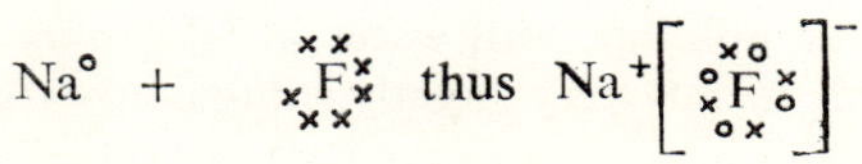

where o and x represent the respective outer shell electrons (this is merely for pictorial clarity: there is no difference between these electrons).

Noble gas structures can also be attained by electron sharing between atoms which do not differ markedly in electronegativity. In methane, for example, molecules are produced, each of which is held together by covalent bonds:

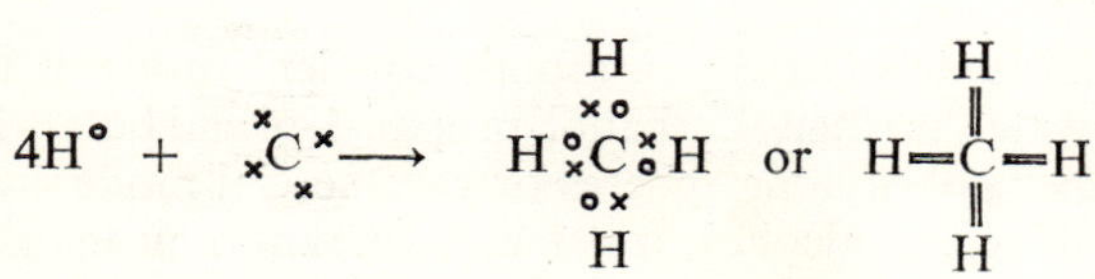

The four covalent bonds of the methane molecule are directed tetrahedrally from the central carbon atom. The covalent bond of a hydrogen molecule consists of an electron pair equally shared between the two hydrogen atoms. In the hydrogen chloride molecule, where the chlorine atom is more electronegative than the hydrogen atom, the electron pair of the covalent bond is more likely to be found nearer the chlorine atom than the hydrogen atom. A small electrical dipole is set up, the magnitude of which depends on the electronegativity difference of the elements concerned. As a result, such a covalent molecule is said to have partial ionic character:

$$\mathrm{H{-}Cl} \quad \text{i.e.} \quad \underset{\delta+}{\mathrm{H}}{-}\underset{\delta-}{\mathrm{Cl}}$$

Variations in the physical properties of substances may be attributed to the predominance of either non-directional electrostatic forces of attraction (i.e. electrovalency) or directional attractive forces which are essentially electrical in character (i.e. covalency).

Questions

1. Why does potassium fluoride have a high melting point, whereas benzene has a low melting point?

2. Why is acetone miscible with water in all proportions, whereas carbon disulphide is only very slightly soluble in water?

3. What cohesive forces exist in liquid bromine?

4. Why do diamond and silica have very high melting points?

12.1 Influence of molecular size on the boiling point, molar volume and refractive index for successive members of a series of normal alcohols*

Requirements

space-filling models (based on atomic covalent radii) of the molecules of water, methanol, ethanol, propan-1-ol and butan-1-ol made from either polystyrene spheres or commercial models

110 cm^3 of each alcohol (methanol, ethanol, propan-1-ol and butan-1-ol) contained in each of four 150 cm^3 stoppered reagent bottles

50 cm^3 liquid paraffin
10 cm^3 acetone
four 100 cm^3 stoppered measuring cylinders
100 cm^3 beaker
four porous chips
hot air blower
refractometer
boiling point apparatus consisting of a 100 cm^3 extraction flask, a 150 mm × 16 mm test tube, a metal stirrer (with a loop to fit round the test tube), a 0°C to 360°C thermometer and appropriately bored corks, assembled as shown:

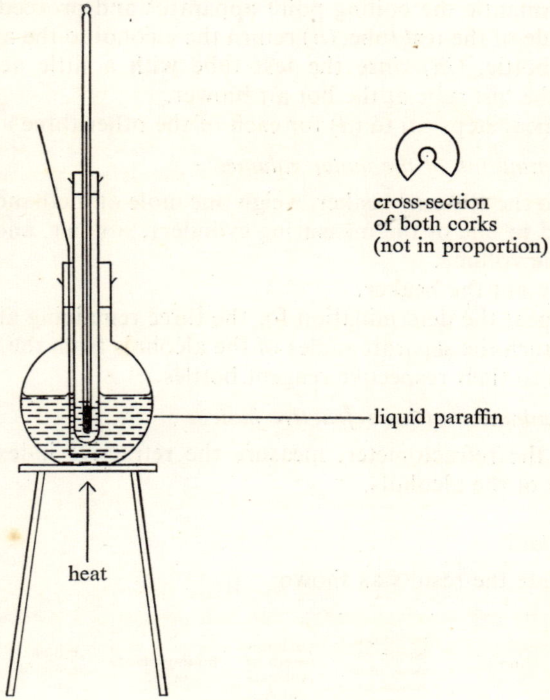

FIG. 26

Procedure

1. Measurement of the molecular models

(*a*) Examine the covalent models of the molecules of water, methanol, ethanol, propan-1-ol and butan-1-ol, and measure the greatest over-all length of each model.

(*b*) Ascertain the scale of the models.

2. Determination of the boiling points

(*a*) Put one ignition tube (c. 4 cm^3) of methanol and a clean dry piece of porous pot in the clean dry test tube of the boiling point apparatus.

(*b*) Assemble the boiling point apparatus, with the bulb of the thermometer dipping into the alcohol but not touching the bottom of the test tube.

(*c*) Gently heat and stir the liquid paraffin until the alcohol boils steadily and the thermometer registers a constant temperature. Record this temperature and immediately stop heating the liquid paraffin.

(*d*) Dismantle the boiling point apparatus and proceed to (*i*) dry the outside of the test tube, (*ii*) return the alcohol to the appropriate reagent bottle, (*iii*) rinse the test tube with a little acetone and (*iv*) dry the test tube at the hot air blower.

(*e*) Repeat steps (*a*) to (*d*) for each of the other three alcohols.

3. Determination of the molar volumes

(*a*) Into the 100 cm^3 beaker, weigh one mole of methanol, transfer the liquid to one of the measuring cylinders, stopper, and read and record the volume.

(*b*) Dry out the beaker.

(*c*) Repeat the determination for the three remaining alcohols.

(*d*) Return the separate moles of the alcohols from the measuring cylinders to their respective reagent bottles.

4. Determination of the refractive indices

Using the refractometer, measure the refractive index of water and each of the alcohols.

Observations

1. Tabulate the results as shown:

water/ alcohol	mole weight	maximum length of molecular model in cm	maximum length of molecule in Å	boiling point in °C	molar volume in cm^3	refractive index

2. Represent the above data by drawing histograms. Plot the value of each of the physical characteristics in the last four columns against the molecular weights of water and the alcohols, and ensure that the axes intersect at zero.

Conclusions

From the histograms, determine, for the addition of each successive $-CH_2$ group, the increase in (*a*) length of the molecule, (*b*) boiling point, (*c*) molar volume and (*d*) refractive index.

Questions

1. (*a*) From the appropriate histograms, what is the expected (*i*) boiling point, (*ii*) molar volume and (*iii*) refractive index of water?

(*b*) What explanation may be given for the variation between the observed and the extrapolated results for water?

2. For the following series of compounds, study and complete the information in the table:

	constitutional formula	name	mole weight	g/cm³ at 20°C	molar volume in cm³
SERIES A	$CH_3.CH_2.CH_2.CH_2OH$		74	0·81	
	$CH_3.CH_2.O.CH_2.CH_3$		74	0·71	
	$CH_3.CH_2.O.CO.CH_3$		88	0·90	
	$CH_3.O.CO.CH_2.CH_3$		88	0·92	
	$CH_3.CO.O.OC.CH_3$		102	1·08	
SERIES B	$CH_3.CH_2OH$		46	0·79	
	$CH_3.CHO$		44	0·78	
	$CH_3.COOH$		60	1·05	

What explanation may be given for the anomalies within Series A and B?

12.2 Hydrogen chloride, HC1*

Requirements

20 cm^3 toluene dried over anhydrous calcium chloride for at least twenty-four hours
10 cm^3 concentrated sulphuric acid
4 cm^3 concentrated hydrochloric acid
2 cm^3 alcoholic silver(I) nitrate solution
2 cm magnesium ribbon
two marble chips
universal indicator paper
four test tubes
teat pipette
electrical circuit for testing the conductivity of solutions
two 150 mm × 20 mm test tubes with side arms, a 4 cm^3 bulb pipette fitted with a 2 cm^3 plastic teat, two right-angled delivery tubes and a thistle funnel which fits into a 100 cm^3 beaker, all assembled, with plastic tubing, as shown:

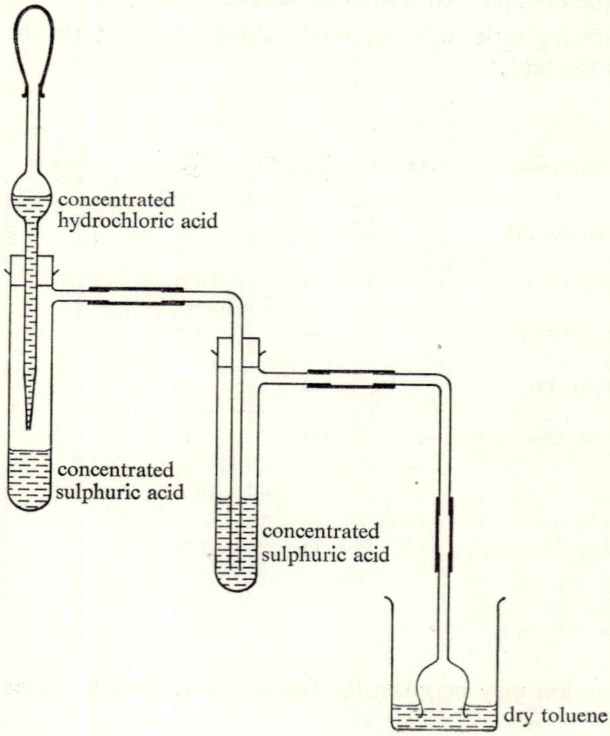

FIG. 27

Procedure

1. Preparation

(*a*) Assemble the dry apparatus.

(*b*) By using the bulb pipette, *cautiously* add concentrated hydrochloric acid dropwise to the concentrated sulphuric acid.

(*c*) Pass a steady stream of the dry gas into the dry toluene for about ten minutes.

2. Examination of the product

Divide the solution of hydrogen chloride in toluene into two equal volumes.

(*a*) Test one of these volumes as follows:

(*i*) dip in a small piece of universal indicator paper,

(*ii*) using the whole volume, determine whether or not the solution conducts an electric current,

(*iii*) divide the solution from step (*ii*) into three portions; to one portion add a marble chip, to another add about 1 cm of magnesium ribbon, and to the third add about 1 cm^3 of the alcoholic silver(I) nitrate solution.

(*b*) To the second volume, add 10 cm^3 of water, shake and allow the layers to separate. With the teat pipette, withdraw all the aqueous layer, and test as in step (*a*).

Tabulate all the observations.

Questions

1. How can the observations from step *2* be explained?

2. What are the equations for all the reactions which have occurred?

12.3 Aluminium chloride, $AlCl_3$*

Requirements

0·5 g clean aluminium turnings
three tablets of potassium permanganate
8 cm^3 concentrated sulphuric acid
4 cm^3 concentrated hydrochloric acid
5 cm^3 carbon tetrachloride (TOXIC : CARE)
1 cm^3 alcoholic silver(I) nitrate solution
1 cm^3 aqueous silver(I) nitrate solution
universal indicator paper
two test tubes
micro burner
electrical circuit for testing the conductivity of a solution
two 150 mm×20 mm test tubes with side arms, a 4 cm^3 bulb

pipette fitted with a 2 cm^3 plastic teat, a right-angled delivery tube, a 150 mm × 12 mm test tube (with a hole blown in the end) and a 125 mm × 16 mm test tube, arranged as shown:

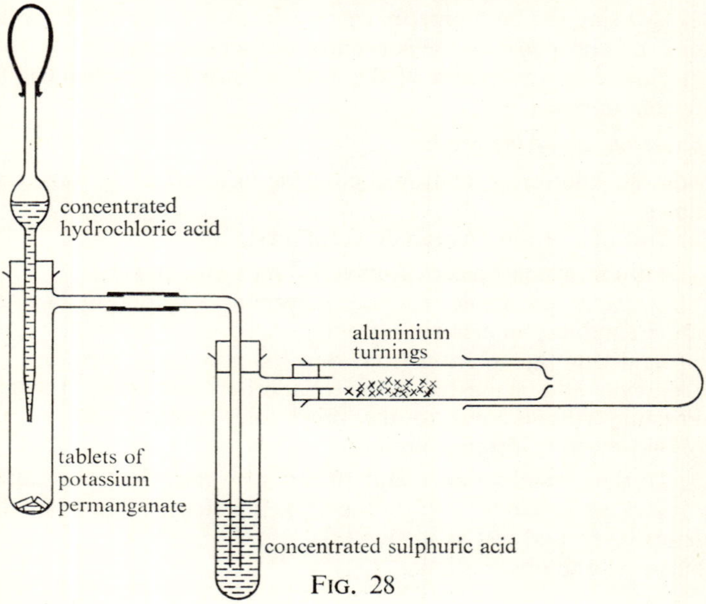

FIG. 28

Procedure

1. Preparation

(*a*) Assemble the dry apparatus.

(*b*) Using the bulb pipette, add a few drops of concentrated hydrochloric acid to the tablets of potassium permanganate to start the evolution of chlorine (heat should *not* be applied).

(*c*) Using a micro burner, gently warm the aluminium turnings in the combustion tube to maintain a steady reaction.

(*d*) Ensure a steady flow of chlorine by adding more concentrated hydrochloric acid when necessary.

2. Examination of the product

(*a*) Dissolve the aluminium chloride in a quarter of a test tube (c. 5 cm^3) of carbon tetrachloride.

(*b*) Determine whether or not this solution conducts an electric current.

(*c*) Divide the solution of aluminium chloride in carbon tetrachloride into three portions and test each, tabulating the observations, as follows:

(*i*) add about 1 cm^3 alcoholic solution of silver(I) nitrate,

(*ii*) add about 1 cm^3 of an aqueous solution of silver(I) nitrate and shake,

(*iii*) add about 1 cm^3 of water, shake and test with universal indicator paper.

Questions

1. What physical property of the aluminium chloride ensures the ease of the preparation?

2. How can the results from part *2* of the experiment be explained?

3. What are the equations for all the reactions which have taken place?

2.4 Iron(II) chloride, $FeCl_2$, and iron(III) chloride, $FeCl_3$

Requirements

200 cm clean iron wire

5·0 g hydrogen chloride generating mixture (an intimate mixture of dry sodium chloride and dry sodium hydrogen sulphate in the ratio 2:1 by weight)

5·0 g chlorine generating mixture (an intimate mixture of dry sodium chloride, dry sodium hydrogen sulphate and dry crushed granular manganese(IV) oxide in the ratio 2:1:1 by weight)

bench reagents: solutions of potassium hexacyanoferrate(II) (i.e. potassium ferrocyanide), potassium hexacyanoferrate(III) (i.e. potassium ferricyanide) and potassium thiocyanate

dry glass wool

match stick

three test tubes

two sets of dry apparatus, each consisting of a 125 mm × 16 mm test tube and a 150 mm × 12 mm rimless test tube, assembled thus:

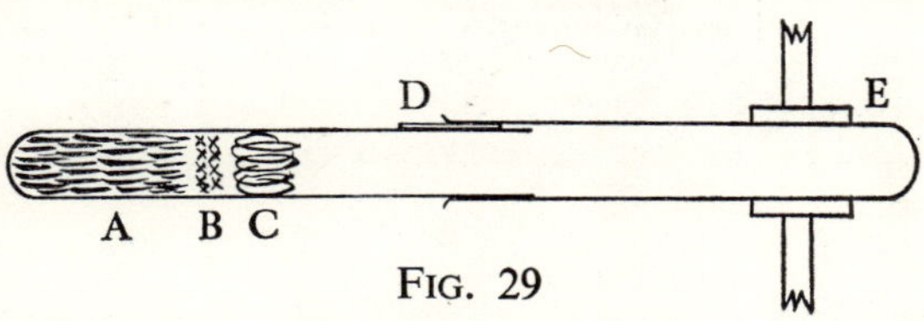

FIG. 29

A mixture for the generation of either hydrogen chloride or chlorine

B glass wool plug

C ball of iron wire

D match stick wedge

E clamp

Procedure

1. Passage of dry hydrogen chloride through heated iron wire

(*a*) Put the hydrogen chloride generating mixture in the rimless test tube to a depth of about 6 cm and cover with a small plug of dry glass wool.

(*b*) Roll about 100 cm of clean iron wire into a small ball and push it down the rimless test tube until it rests on the plug of glass wool.

(*c*) Assemble the two test tubes as illustrated and hold horizontally in the clamp.

(*d*) *Strongly* heat the iron wire from all sides with a Bunsen burner flame to attain red heat, and continue this heating with intermittent heating of the hydrogen chloride generating mixture.

(*e*) When all signs of reaction have ceased, remove the condenser test tube and pour into it an ignition tube (c. 4 cm^3) of water. Divide the resulting solution into three equal volumes and test it by dropwise addition of solutions of (*i*) potassium hexacyanoferrate(II), (*ii*) potassium hexacyanoferrate(III) and (*iii*) potassium thiocyanate.

2. Passage of dry chlorine through heated iron wire

Repeat the same procedure, but employ a 6 cm depth of the mixture for the generation of chlorine instead of that for hydrogen chloride.

Observations

Tabulate the results thus:

	potassium hexacyanoferrate(II)	potassium hexacyanoferrate(III)	potassium thiocyanate
iron + hydrogen chloride			
iron + chlorine			

Questions

1. What evidence is there that the products of the two formation reactions are covalent?

2. What are (*a*) the molecular equations for the formation reactions and (*b*) the ionic equations for the test reactions?

12.5 Tin(IV) chloride, $SnCl_4$*

Requirements

1·0 g granulated tin
three tablets of potassium permanganate
10 cm^3 concentrated sulphuric acid
5 cm^3 concentrated hydrochloric acid
3 cm^3 carbon tetrachloride (TOXIC:CARE)
10 cm^3 bench ammonium hydroxide
7·5 cm diameter watch glass
two test tubes
two 150 mm × 20 mm test tubes with side arms, a 4 cm^3 bulb pipette fitted with a 2 cm^3 plastic teat, a right-angled delivery tube, a 150 mm × 12 mm test tube (with a hole blown in the end), a condenser adapter with a wide end of 20 mm internal diameter, a 125 mm × 16 mm test tube and a 250 cm^3 plastic carton (cut from a detergent bottle), set up as shown:

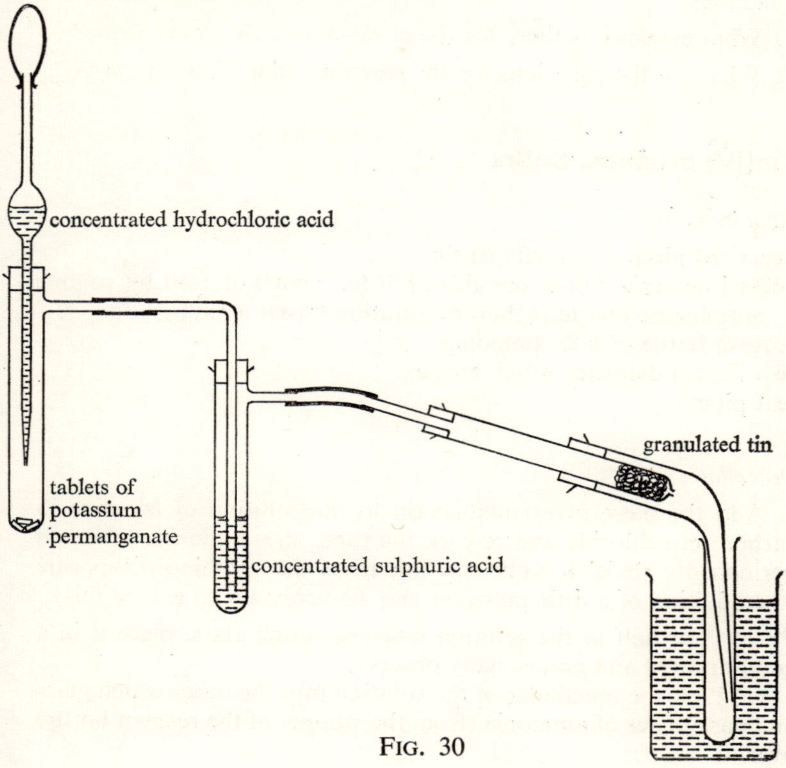

FIG. 30

Procedure

1. Assemble the dry apparatus with about a 2 cm depth of granulated tin in the reaction test tube.

2. From the bulb pipette, add drops of concentrated hydrochloric acid to the tablets of potassium permanganate to start and maintain a steady stream of chlorine (do *not* heat).

3. When all signs of reaction have ceased, withdraw the collecting tube and dry the outside of it.

4. Pour half of the condensed liquid into a dry test tube containing the carbon tetrachloride, shake and observe. Pour the liquid onto the watch glass, set aside, and examine from time to time.

5. Slowly add the remaining half of the condensed liquid to half a test tube of the bench ammonium hydroxide and observe.

Questions

1. What evidence is there for the covalency of tin(IV) chloride?

2. What are the equations for the reactions which have occurred?

12.6 Tin(IV) bromide, $SnBr_4$

Requirements

pea-sized piece of granulated tin
corked test tube about one-third full (c. 7 cm^3) of 1:10 by volume bromine/carbon tetrachloride solution (TOXIC:CARE)
reagent bottle of 0·88 ammonia
two 7·5 cm diameter watch glasses
teat pipette

Procedure

1. Add the piece of granulated tin to the solution of bromine in carbon tetrachloride and re-cork the tube. Set aside the tube and periodically shake it until the colour of the bromine disappears (the addition of a little more tin may be necessary to ensure this).

2. (*a*) Pour half of the solution into one watch glass, place it in a good draught and occasionally observe.

(*b*) Pour the remainder of the solution into the other watch glass and pass fumes of ammonia (from the stopper of the reagent bottle) over it.

Questions

1. What evidence is there for tin(IV) bromide being covalent?
2. What are the equations for the reactions which have taken place?

12.7 Disulphur dichloride, S_2Cl_2*

Requirements

2·0 g flowers of sulphur
7·0 g fused calcium chloride
three tablets of potassium permanganate
5 cm^3 concentrated hydrochloric acid
3 cm^3 benzene (TOXIC:CARE)
small plug of glass wool
ignition tube
two test tubes
micro burner
one 150 mm × 20 mm test tube with side arm, a 4 cm^3 bulb pipette fitted with a 2 cm^3 plastic teat, a 150 mm calcium chloride tube, a 150 mm × 12 mm test tube (with a hole blown in the end), a condenser adapter with a wide end of 20 mm internal diameter, a 125 mm × 16 mm test tube and a 250 cm^3 plastic carton (cut from a detergent bottle), set up as shown:

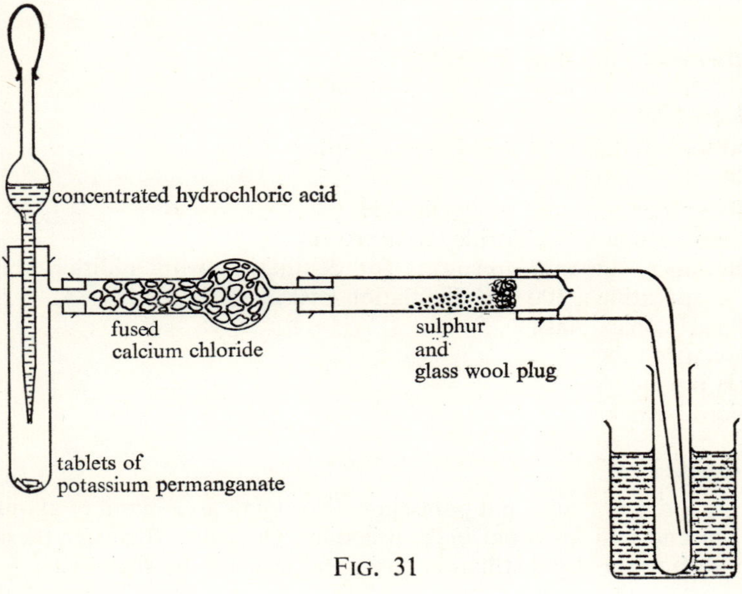

FIG. 31

Procedure

1. Assemble the dry apparatus, putting the small plug of glass wool followed by half an ignition tube (c. 2 g) of sulphur in the reaction tube.

2. From the bulb pipette, add drops of concentrated hydrochloric acid to the tablets of potassium permanganate to start and maintain a steady stream of chlorine (do *not* heat).

3. When all the air has been displaced by chlorine, heat the flowers of sulphur gently with the micro burner until molten. Continue to pass the chlorine over the molten sulphur to obtain at least 1 cm^3 of sulphur monochloride in the collecting tube.

4. Put about one-third of the sulphur monochloride in a test tube, add about half an ignition tube (c. 2 cm^3) of benzene, shake and observe.

5. To the remaining sulphur monochloride, add about 5 cm^3 of purified water and warm. Identify the three hydrolysis products.

Questions

1. What is the chief impurity of sulphur monochloride, and how can it be easily removed?

2. What are the equations for all the reactions which have occurred?

12.8 Chromyl chloride, CrO_2Cl_2

Requirements

4·0 g potassium dichromate crystals (fine)
4·0 g sodium chloride
10 cm^3 concentrated sulphuric acid
6 cm^3 carbon tetrachloride (TOXIC:CARE)
semi-micro organic apparatus for distillation with addition (incorporating a 100 cm^3 distillation flask)
50 cm^3 conical flask
three test tubes
teat pipette
micro burner

Procedure

1. In a dry test tube, put potassium dichromate to a depth of about 2 cm and add an equal bulk of sodium chloride. Transfer these solids into the dry distillation flasks and mix well by shaking.

2. Assemble the dry semi-micro organic apparatus for distillation with addition, employing the 50 cm^3 conical flask as the collecting vessel.

3. Put the concentrated sulphuric acid into the dropping funnel and add about 1 cm^3 of it to the mixture. When the reaction has almost ceased, add a further 1 cm^3 of the acid. Continue in this manner until all the acid has been added.

4. Heat the distillation flask *gently*, to avoid undue frothing, with the micro burner flame, and distil the chromyl chloride into the conical flask.

5. Employing the teat pipette, add a few drops of the product to a dry test tube a quarter full (c. 5 cm^3) of carbon tetrachloride and observe.

6. Half fill a test tube with water and, by dropwise addition from the teat pipette, add the remaining chromyl chloride; test the resulting solution for the presence of chromate and chloride ions.

Questions

1. What evidence is there that chromyl chloride is covalent?

2. What is the equation for the reaction between potassium dichromate, sodium chloride and concentrated sulphuric acid?

3. What type of reaction takes place in step *6* above, and what are the equations for this reaction and the subsequent test reactions employed?

4. Chromyl bromide is unstable and chromyl iodide is unknown. How can this be explained?

13

HYDROGEN BONDING

Introduction

In chemical combination a hydrogen atom normally forms only one predominantly covalent link, but in certain circumstances it can link two atoms of markedly electronegative elements. The fluorine atom in the hydrogen fluoride molecule attracts the electrons of the covalent bond, and the molecule is therefore polarised:

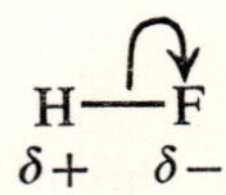

The positively charged hydrogen atom is then attracted to the fluorine atom of another molecule, and a weak hydrogen bond results:

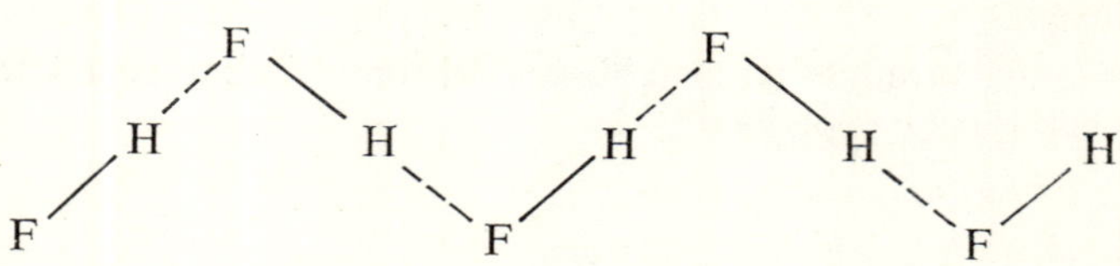

The anomalous physical properties of water, including the open hexagonal structure of ice, arise from intermolecular hydrogen bonding, which causes association between water molecules:

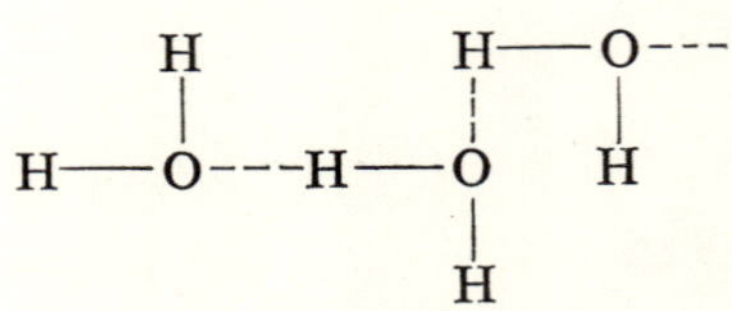

The success of winter sports and the glacial attractions of the Swiss Alps depend upon the existence of the hydrogen bond. The correct chopping of wood is determined by hydrogen bonds which break along the grain as opposed to covalent bonds which have to be broken by chopping across the grain. The reason why biochemical reactions take place so easily at room temperature is also attributed to hydrogen bonding.

Questions

1. Why is the density of water a maximum at 4°C?

2. What are the relative strengths of an average hydrogen bond compared with a typical covalent bond?

3. What is the structural arrangement for hydrogen bonding in an aqueous solution of ammonia (show by diagram)?

4. 2-hydroxybenzoic acid exhibits intramolecular hydrogen bonding: what is this?

3.1 Evidence for the existence of the hydrogen bond by mixing organic liquids

Theory

The carbon tetrachloride / acetone, chloroform / acetone and chloroform/carbon tetrachloride systems may be investigated by experimentally determining the volume and thermal changes on mixing equimolar volumes of the constituent liquids of each system.

Data

Record and complete the following table:

liquid	mole weight	g/cm³ at 20°C	molar volume in cm³ at 20°C	specific heat in cal/g°C
carbon tetrachloride	154·0	1·59		0·20
acetone	58·0	0·79		0·53
chloroform	119·5	1·48		0·23

On completion of this data table, it will be appreciated that half molar volumes of each liquid will be suitable for both investigations.

A Determination of the volume change on mixing

Requirements

120 cm³ carbon tetrachloride (TOXIC : CARE)
150 cm³ acetone
100 cm³ chloroform
50 cm³ industrial alcohol
two 50 cm³ burettes (plastic tube/pinch-clip type)
100 cm³ graduated tube
filter funnel
hot air blower

Procedure

1. Wash out the burettes, graduated tube and filter funnel with water, a little alcohol and then a little acetone, and dry them at the blower.

2. Clamp the graduated tube vertically below bench level, with the filter funnel resting in its mouth.

3. Fill one burette with carbon tetrachloride and the other with acetone.

4. Clamp the burettes so that (*a*) they form a 'Y' with the graduated tube and (*b*) their jets are adjacent to each other within the filter funnel.

5. Simultaneously run half molar volumes of each liquid together, to ensure thorough mixing, into the filter funnel. Record the final volume of the mixed liquids in the graduated tube. Pour the mixed liquids into a suitable residue bottle.

6. Repeat steps *1* to *5* for the chloroform/acetone system and then for the chloroform/carbon tetrachloride system.

Observations

Record the volume data in tabular form thus:

system	half molar volumes in cm³ of constituent liquids before mixing		total volume in cm³ before mixing	volume in cm³ after mixing
carbon tetrachloride/acetone				
chloroform/acetone				
chloroform/carbon tetrachloride				

Questions

1. How can the volume observations be fully explained?

2. The heats of mixing of equimolar portions of individual members of the normal alcohol series, $CnH_{(2n+1)}$ OH, with a fixed volume of water decrease with an increase in *n*. Why is this?

B *Determination of the heat of mixing*

Requirements

120 cm^3 carbon tetrachloride (TOXIC : CARE)
200 cm^3 acetone
100 cm^3 chloroform
50 cm^3 industrial alcohol
two 50 cm^3 burettes (plastic tube/pinch-clip type)
two -10°C to 110°C thermometers
100 cm^3 plastic bottle
vacuum flask fitted with one of the -10°C to 110°C thermometers through a plastic lid and ring of rubber tubing†
hot air blower

Procedure

1. Wash out the burettes, plastic bottle and vacuum flask with water, a little alcohol and then a little acetone, and dry them at the blower.

2. Put carbon tetrachloride into one burette and acetone into the other. Run a half molar volume of (*a*) carbon tetrachloride into the plastic bottle and (*b*) acetone into the vacuum flask.

3. Insert a dry thermometer into each of the two liquids and take their steady temperatures. Immediately pour the carbon tetrachloride into the vacuum flask. Stir well with the thermometer and then record the steady temperature attained after thorough mixing.

4. Pour the mixed liquids into a suitable residue bottle and dry the thermometers.

5. Repeat steps *1* to *4* for the chloroform/acetone system (again with the acetone initially in the vacuum flask) and then for the chloroform/carbon tetrachloride system.

† As set up in experiment 3.3.

Observations

1. Record for the carbon tetrachloride/acetone system:

Initial temperature of carbon tetrachloride = °C
Initial temperature of acetone = °C
Steady temperature after mixing = °C

2. Repeat similarly for the other two systems.

Calculation

Make allowance for the thermal capacity of the vacuum flask and fitments† and calculate the heat of mixing for each system, expressing the heat quantities in kilocalories per mole.

Question

How can the observed heat quantities be fully explained?

13.2 Investigation of the distribution of acetic acid between water and carbon tetrachloride*

Theory

If acetic acid dissolves in the two immiscible solvents water and carbon tetrachloride, such that in water it is present as simple molecules, CH_3COOH, and in carbon tetrachloride as associated molecules, $(CH_3COOH)_n$, then for the equilibrium

$$\frac{(\text{concentration of acetic acid in water})^n}{\text{concentration of acetic acid in carbon tetrachloride}} = \text{constant}$$

i.e.

$$\frac{[CH_3COOH]^n_{H_2O}}{[(CH_3COOH)_n]_{CCl_4}} = K$$

$$n \log[CH_3COOH]_{H_2O} - \log[(CH_3COOH)_n]_{CCl_4} = \log K.$$

On plotting $\log[(CH_3COOH)_n]_{CCl_4}$ against $\log[CH_3COOH]_{H_2O}$, the value of n can be obtained from the slope of the graph.

Requirements

120 cm^3 1·0M sodium hydroxide solution
60 cm^3 0·05M sodium hydroxide solution

† As determined in experiment 3.3.

60 cm^3 c. 2·5M acetic acid
120 cm^3 carbon tetrachloride (TOXIC : CARE)
phenolphthalein indicator solution
specified volumetric ware plus a 10 cm^3 pipette, a 50 cm^3 pipette and a 50 cm^3 burette
250 cm^3 reagent bottle with glass stopper to fit
two adhesive labels
laboratory tissues

Procedure

1. Fill a communal burette (plastic tube/pinch-clip type) with carbon tetrachloride, label appropriately and then run 50 cm^3 of carbon tetrachloride into the reagent bottle.

2. Attach the pipette filler to the 50 cm^3 pipette and transfer a 50 cm^3 portion of the acetic acid to the reagent bottle.

3. Stopper the reagent bottle, shake well over a period of at least ten minutes and then set aside for the layers to separate.

4. Fill the two individual burettes with the 1·0M and 0·05M sodium hydroxide solutions respectively, and again label appropriately.

5. Attach the pipette filler to the 10 cm^3 pipette.

6. Unstopper the reagent bottle, lower the tip of the 10 cm^3 pipette into the upper water layer and proceed to withdraw a 10 cm^3 sample. Transfer this aqueous portion to a conical flask and thoroughly wash out the pipette with water.

7. Lower the tip of the 10 cm^3 pipette into the carbon tetrachloride layer and gently blow out any of the aqueous layer which has entered the stem. Withdraw a 10 cm^3 portion of the non-aqueous layer and with a laboratory tissue dry the lower stem of the pipette. Transfer the portion to the second conical flask, thoroughly wash out the pipette with water and well flush out the pipette filler with air.

8. To the contents of the reagent bottle, add 10 cm^3 of carbon tetrachloride from the labelled burette, and a 10 cm^3 pipetted portion of water. Stopper and shake the bottle over a period of at least five minutes.

9. Titrate the 10 cm^3 aqueous portion with the 1·0M sodium hydroxide solution and employ phenolphthalein as indicator.

10. Titrate the 10 cm^3 non-aqueous portion with the 0·05M sodium hydroxide solution, again employ phenolphthalein as indicator and shake the contents of the flask thoroughly after each addition of alkali.

11. Repeat steps *6* to *10* to obtain at least five pairs of titration results.

Observations and Calculations

1. Complete a table headed as follows:

cm^3 of 1·0M NaOH for 10 cm^3 of H_2O layer	cm^3 of 0·05M NaOH for 10 cm^3 of CCl_4 layer	concentration in mole/l of acetic acid in		log $[(CH_3COOH)]_{H_2O}$	log $[(CH_3COOH)_n]_{CCl_4}$
		H_2O layer $[CH_3COOH]_{H_2O}$	CCl_4 layer $[(CH_3COOH)_n]_{CCl_4}$		

2. Plot $\log[(CH_3COOH)_n]_{CCl_4}$ against $\log[CH_3COOH]_{H_2O}$ and ascertain the values of *n* and *K*.

Questions

1. What explanation can be given for the value of *n* obtained?

2. How and why will the distribution between water and carbon tetrachloride of the normal carboxylic acids, $C_nH_{(2n+1)}$ COOH, vary with mole weight?

3. (*a*) How may the expression

$$\frac{[CH_3COOH]^n_{H_2O}}{[(CH_3COOH)_n]_{CCl_4}} = K$$

be derived by application of the Laws of Mass Action and Distribution to the equilibria shown in the accompanying diagram?

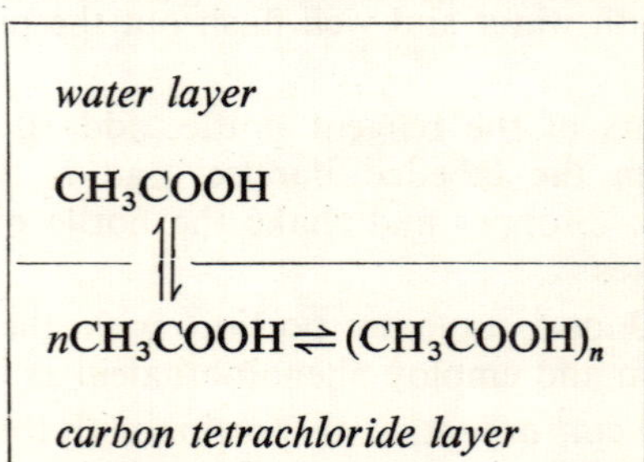

(*b*) What is the true state of the acetic acid in the aqueous layer?

13.3 Cryoscopic investigation of the association of benzoic acid in benzene

Theory

The mole weight of a solute in benzene can be determined from the following relationship† :

$$\Delta\theta = 5{\cdot}1 \times \frac{c}{M}.$$

Benzoic acid dissolved in benzene exists as an equilibrium of single and associated molecules which are the outcome of hydrogen bonding. Considering one mole of benzoic acid, then

$$nC_6H_5COOH \longrightarrow (C_6H_5COOH)_n$$

if moles initially are	1	–
and moles at equilibrium are	$1-\alpha$	$\frac{\alpha}{n}$

where α is the degree of association.

For a fixed mass of solute, the number of particles dissolved is inversely proportional to the mole weight of the solute. Thus, if

M or $M_{obs.}$ = mole weight observed by experiment,

and $M_{calc.}$ = mole weight of single molecules calculated from the appropriate atomic masses,

then
$$\frac{M_{calc.}}{M_{obs.}} = \frac{1-\alpha+\alpha/n}{1}.$$

However, assuming that benzoic acid dimerizes in benzene, $n = 2$, then

$$\frac{M_{calc.}}{M_{obs.}} = \frac{2-\alpha}{2}$$

$$\alpha = \frac{2(M_{obs.}-M_{calc.})}{M_{obs.}}$$

where α is a measure of the degree of hydrogen bonding.

Requirements

1·5 g A.R. benzoic acid
30 cm^3 benzene (TOXIC: CARE)
ice
50 cm^3 burette
-10°C to 110°C thermometer
stop-clock

† See experiment 5.1.

hand lens

four pieces of glazed paper (c. 8 cm×8 cm)

350 cm^3 plastic carton and lid, boiling tube and a two-holed cork, a -5°C to 50°C (in 0·1°C) thermometer and a wire stirrer, assembled as in figure 8 on page 67.

Procedure

Carry out as in experiment 5.1, with the following exceptions:

(*a*) adjust the temperature of the water to about 1°C in steps *1* and *5*,

(*b*) note that the benzoic acid requires no drying in step *6*,

(*c*) employ 0·25 g portions of benzoic acid throughout.

Observations and Calculations

1. Draw up and complete a table with the following headings:

cm^3 of benzene	g of benzene (density = 0·88 g/cm^3 at 20°C)	g of benzoic acid	g of benzoic acid per 1000 g of benzene	freezing point in °C	depression of freezing point in °C	$5{\cdot}1\times\frac{c}{\Delta\theta}$	$M_{calc.}$	$\frac{2(M_{obs.}-M_{calc.})}{M_{obs.}}$
			c			$M_{obs.}$		α

2. On the same axes, plot graphs of (*a*) observed mole weight of solute, $M_{obs.}$, and (*b*) degree of association, α, against concentration, c.

Questions

1. What is the significance of 5·1 in the formula $\Delta\theta = 5{\cdot}1\times\frac{c}{M}$?

2. How can the forms of the graphs be explained?

3. What is the condition of benzoic acid when dissolved in water, acetone, glacial acetic acid and carbon tetrachloride respectively?

4. Why is the observed mole weight of salicylic acid dissolved in benzene approximately equal to the calculated mole weight?

14

CONDUCTIVITY

Introduction

The conductivity of an electrolyte in aqueous solution depends on the rate of movement of the constituent ions, i.e. upon their mobilities, and on the concentration of these ions.

The mobilities of ions under similar conditions differ according to their charge and their size. Ions with multiple charges tend to move more quickly than ions with a single charge: the former are more strongly attracted by oppositely charged electrodes. Small ions tend to have higher mobilities than large ions because their passage through the electrolyte media is less impeded.

Questions

1. What is the definition of the absolute mobility of an ion?

2. The absolute mobilities of hydrogen and hydroxyl ions are very high compared with those for other ions. How may this be fully explained?

3. The relative absolute mobilities at 25°C of the nitrate, chloride, bromide and sulphate ions are 1·00:1·07:1·10:1·12. What explanations may be given for these figures?

Notes on apparatus

A 'specified conductivity apparatus' is required for all the experiments on conductivity; this is basically as shown in figure 32 on page 204.

The equipment, obtainable from Griffin and George Ltd., Frederick St., Birmingham 1, is not designed for measuring actual conductivities, but it provides evidence for the presence of ions in solution. The equipment and the experiments are based on the ideas of Mr. A. H. Johnstone, B.Sc., A.R.I.C.

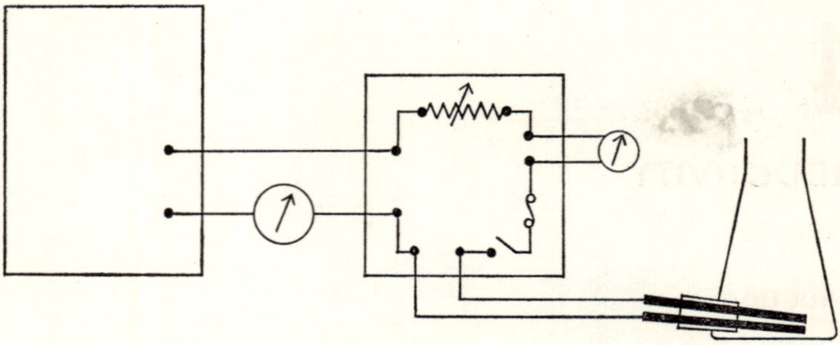

FIG. 32

14.1 Comparison of the conductivities of equimolar aqueous solutions of binary electrolytes

Requirements

60 cm^3 0·1M hydrochloric acid
60 cm^3 0·1M sodium hydroxide solution
60 cm^3 0·1M sodium chloride solution
specified conductivity apparatus
boiling tube

Procedure

1. Set up the conductivity apparatus.

2. (*a*) Pour a boiling tube (c. 55 cm^3) of the hydrochloric acid into the titration flask.

(*b*) Switch on and adjust the LT variable voltage supply and/or rheostat to give a maximum reading on the 100 mA meter. Switch off the cell.

(*c*) Switch on the cell, immediately record the ammeter reading, and switch off the cell.

(*d*) Wash out the titration flask with purified water.

3. Repeat step *2* using a solution of (*a*) sodium hydroxide and (*b*) sodium chloride.

Questions

1. How can the differences in the conductivities of the three electrolytes be explained?

2. At 25°C the molar conductivities at infinite dilution for aqueous solutions of the chlorides of lithium, sodium and potassium are 115, 126 and 150 $ohm^{-1}.cm^2$ respectively. How can these figures be explained?

14.2 Effect of dilution on the conductivity of aqueous solutions of electrolytes

Requirements

50 cm^3 concentrated hydrochloric acid
50 cm^3 glacial acetic acid
50 cm^3 0·88 ammonia
specified conductivity apparatus
test tube
adhesive label

Procedure

1. Set up the conductivity apparatus.

2. Stick the adhesive label to the test tube so that the top edge is at the middle of the test tube.

3. (*a*) Pour two test tubes (c. 40 cm^3) of the concentrated hydrochloric acid into the titration flask.

(*b*) Switch on and adjust the LT variable voltage supply and/or rheostat to give a maximum reading on the 100 mA meter. Switch off the cell.

(*c*) Switch on the cell, immediately record the ammeter reading, and switch off the cell.

(*d*) Add half a test tube (c. 10 cm^3) of purified water to the hydrochloric acid, swirl gently and repeat step (*c*).

(*e*) Repeat step (*d*) until the titration flask is full.

4. Repeat step *3* for the glacial acetic acid and then the 0·88 ammonia.

Observations

Tabulate the observations as shown on page 206 and express graphically.

c. 10 cm^3 portions water added		0	1	2	3	4
meter reading in mA for:	HCl					
	CH_3COOH					
	NH_4OH					

Interpretation

Explain the variations implicit in the graphs.

4.3 Conductivity changes for various titrations

Requirements (for each of the four experiments)

2 cm^3 acid
5 cm^3 alkali
indicator solution
specified conductivity apparatus
boiling tube
teat pipette

A Titration of a strong acid with a strong base

Employ c. 1·0M sulphuric acid and c. 1·0M sodium hydroxide solution with bromothymol blue as indicator.

Procedure

1. Set up the conductivity apparatus.

2. Pour a boiling tube (c. 55 cm^3) of purified water into the titration flask and add ten drops of the sulphuric acid from the teat pipette. Add three drops of the indicator and swirl gently.

3. Switch on and adjust the LT variable voltage supply and/or rheostat to give a maximum reading on the ammeter. Switch off the cell.

4. Switch on the cell, immediately record the ammeter reading, and switch off the cell.

5. Wash out the teat pipette.

6. From the teat pipette, add five drops of the sodium hydroxide solution, swirl and repeat step *4*.

7. Repeat step *6* for another seven, separate five-drop additions of alkali, noting the addition which causes the indicator to change colour.

Observations

Tabulate as follows:

drops of c. 1·0M NaOH added to ten drops of c. 1·0M H_2SO_4	0	5	10	15
meter reading in mA				

and express the information graphically.

Interpretation

Explain the form of the graph and state whether it confirms the indicator colour change.

B Titration of a strong acid with a weak base

Repeat experiment A using ten drops of c. 2·0M hydrochloric acid and a total addition of forty drops of c. 1·0M ammonium hydroxide solution with methyl red as indicator.

C Titration of a weak acid with a strong base

Repeat experiment A using ten drops of c. 2·0M acetic acid and a total addition of forty drops of c. 1·0M sodium hydroxide solution with phenolphthalein as indicator.

D Titration of a weak acid with a weak base

Repeat experiment A using ten drops of c. 2·0M acetic acid and a total addition of forty drops of c. 1·0M ammonium hydroxide solution with neutral red screened by methylene blue as indicator.

14.4 Conductivity changes for the neutralisation of an acid by a slightly soluble base

Requirements

1·0 g barium hydroxide
1·0 g calcium hydroxide
4 cm^3 c. 2·0M sulphuric acid
specified conductivity apparatus

ignition tube
boiling tube
teat pipette
stop-clock

Procedure

1. Set up the conductivity apparatus.

2. (*a*) Pour a boiling tube (c. 55 cm^3) of purified water into the titration flask and add ten drops of the sulphuric acid.

(*b*) Switch on and adjust the LT variable voltage supply and/or rheostat to give a maximum reading on the 100 mA meter. Switch off the cell.

(*c*) Switch on the cell, immediately record the ammeter reading, and switch off the cell.

(*d*) Switch on the cell and simultaneously add half an ignition tube of the barium hydroxide (c. 1·0 g) and start the clock; swirl continuously and record the ammeter reading every ten seconds until the readings show no variation.

3. Repeat step *2* employing the calcium hydroxide in place of the barium hydroxide.

Observations

Tabulate the readings as follows:

time in sec		0	10	20	30
meter readings in mA	$Ba(OH)_2$				
	$Ca(OH)_2$				

and express the information graphically.

Interpretation

Explain the variations in the graphs.

14.5 Conductivity changes for a precipitation reaction

Requirements

2 cm^3 c. 0·2M chloride solution
10 cm^3 c. 0·1M silver(I) nitrate solution
specified conductivity apparatus
boiling tube
teat pipette

Procedure

Repeat steps *1* to *7* of experiment 14.3A, employing ten drops of the sodium chloride solution in the titration flask with addition of separate five-drop portions of the silver(I) nitrate solution (ignore instructions concerning the indicator).

Observations

Tabulate as follows:

drops of c. 0·1M $AgNO_3$ added to ten drops of c. 0·2M NaCl	0	5	10	15
meter reading in mA				

and express the figures graphically.

Interpretation

Explain the shape of the graph.

14.6 Conductivity investigation of buffer action*

Requirements

100 cm³ acetic acid/sodium acetate buffer of pH 4
20 cm³ c. 1·0M nitric acid
20 cm³ c. 1·0M sodium hydroxide solution
specific conductivity apparatus
test tube
teat pipette

Procedure

1. Set up the conductivity apparatus.

2. (*a*) Pour two test tubes (c. 40 cm³) of purified water into the titration flask.

(*b*) Switch on and adjust the LT variable voltage supply and/or rheostat to give an ammeter reading of 10 mA. Switch off the cell.

(*c*) Add three drops of the nitric acid and swirl; switch on the cell, immediately record the ammeter reading, and switch off the cell.

(*d*) Repeat step (*c*) to give twelve readings in all.
(*e*) Wash out the titration flask.

3. Repeat step *2* with addition of the sodium hydroxide solution.

4. Repeat steps *2* and *3* employing the buffer solution in place of purified water at the outset.

Observations

For the nitric acid additions, tabulate as follows:

drops of c. 1·0M HNO_3 added	0	3	6	9
water meter reading in mA				
buffer meter reading in mA				

and present the sodium hydroxide information similarly.

Plot the data from the two tables.

Interpretation

1. Explain the different conductivity curves recorded for the water and buffer solution. This explanation should be based upon a consideration of ionic equilibria.

2. The expression

$$[H^+] = K_a \frac{[\text{acid}]}{[\text{acetate}]}$$

is true for an acetic acid/sodium acetate buffer where K_a is the dissociation constant of acetic acid.

(*a*) Show how this expression is derived from ionic equilibria existing in the buffer solution.

(*b*) From the above expression, derive an equation for the pH of the buffer solution.

3. Explain why a mixture of ammonia and ammonium chloride in aqueous solution is a buffer.

15

INDICATORS

Introduction

Ostwald realised that the most commonly used indicators were weak acids, and in 1891 he put forward a simple theory of their action. The ionisation of a weak acid indicator, HIn, is

$$\underset{\text{colour A}}{\text{HIn}} \rightleftharpoons \underset{\text{colour B}}{\text{H}^+ + \text{In}^-}.$$

In acid solution the excess of hydrogen ions would drive the equation to the left, and thus colour A of the undissociated molecules would predominate. Colour B of the indicator anion would show up with the addition of hydroxyl ions; neutralisation of hydrogen ions occurs and shifts the equilibrium to the right.

The equilibrium constant for the weak acid indicator is

$$K_i = \frac{[H^+][In^-]}{[HIn]}$$

where K_i is known as the ionisation constant of the indicator. When the indicator colour change is at the halfway point between the extremes of colour, it may be assumed that there will be equal concentrations of indicator molecules and indicator anions, i.e. $[HIn] = [In^-]$. At this halfway point, K_i will be equal to $[H^+]$, the pH for which is known as the *pK* value for the indicator. Indicators of different *pK* values may be selected for work at varying values of pH.

Questions

1. What is the effect of dilution on the colour of an indicator? Base the answer on the simple Ostwald theory given above.

2. The action of 4-nitrophenol as an acid-base indicator cannot be explained by Ostwald's theory. What is its mode of action? Explain by reference to the *modern theory* of indicators which may be found in textbooks of physical chemistry.

15.1 Indicator charts

Introduction

The pH of a solution is defined as the negative logarithm to the base 10 of the hydrogen ion concentration expressed in moles per litre. Thus an acid solution containing 1 mole per litre of hydrogen ions, i.e. $[H^+] = 1$, will have a pH of 0, and one containing 0·1 mole per litre, i.e. $[H^+] = 0{\cdot}1$, will have a pH of 1.

At 25°C the ionic product of water, $[H^+][OH^-] = 10^{-14}$, so the hydrogen ion concentration can be represented by

$$[H^+] = \frac{10^{-14}}{[OH^-]}.$$

In an alkaline solution, where $[OH^-] = 1$, the hydrogen ion concentration $[H^+] = 10^{-14}$, and this represents a pH of 14. Similarly, $[OH^-] = 10^{-1}$ corresponds to $[H^+] = 10^{-13}$, and a pH of 13.

A *Methyl violet indicator chart by dilution*

Requirements

10 cm^3 1·0M hydrochloric acid
methyl violet indicator solution
1 cm^3 pipette
100 cm^3 measuring cylinder
1000 cm^3 measuring cylinder
coloured pencils

Procedure

1. Use the coloured pencils to complete the following chart for the continuous colour gradation observed on dilution of methyl violet:

pH	0	1	2	3
colour shading				

2. Put six drops of methyl violet into the 100 cm^3 measuring cylinder and add, by pipette, 1 cm^3 of the 1·0M hydrochloric acid (pH = 0).

3. (*a*) Slowly add purified water to (*i*) the 10 cm^3 mark (pH = 1), (*ii*) the 30 cm^3 mark (pH = 1·5) and (*iii*) the 100 cm^3 mark (pH = 2).

(*b*) Transfer the solution to the larger measuring cylinder and slowly dilute to 1000 cm^3 (pH = 3).

Question

What bearing have these observations on the satisfactory detection of end points in acid-base titrations?

B Construction of indicator charts†*

Requirements

25 cm³ 1·0M hydrochloric acid
25 cm³ 1·0M sodium hydroxide solution
indicators covering the pH range‡
10 cm³ pipette
100 cm³ graduated flask
100 cm³ beaker
coloured pencils
test tube block containing fifteen similar test tubes labelled pH 0 to pH 14.

Procedure

1. Two-thirds fill the first test tube with 1·0M hydrochloric acid. This gives a solution of pH = 0.
2. Pipette 10 cm³ of the 1·0M hydrochloric acid into the 100 cm³ graduated flask, make up to the mark with purified water and shake well. This gives a solution of pH = 1 (0·1M).
3. Transfer the contents of the graduated flask to the beaker. Wash out the second test tube with some of the solution in the beaker and then two-thirds fill this test tube with the solution of pH = 1. Wash out the graduated flask with purified water. Pipette 10 cm³ of the 0·1M solution (pH = 1) into the flask, dilute to the mark with purified water and shake well to give a solution of pH = 2.

† This is best performed as a class experiment where each person uses a different indicator.

‡ Employ either Series I or Series II and B.D.H. universal indicator, pH range 4 to 11.

Series I	*pH change*	*Series II*	*pH change*
Methyl violet	0·1- 2·0	Haemotoxylin	0·0- 1·0
Martius yellow	2·0- 3·2	Disulphine blue	0·8- 3·0
Bromophenol blue	2·8- 4·6	Congo red	3·0- 5·0
Methyl red	4·2- 6·3	4-nitrophenol	5·0- 7·0
Bromothymol blue	6·0- 7·6	Phenol red	6·6- 8·4
Phenol red	6·6- 8·4	Phenolthymolphthalein	8·4-11·0
Phenolphthalein	8·4-10·0	Poirrier's blue	11·0-13·0
Alizarin yellow	10·0-12·0		
Titan yellow	12·0-13·0		

H*

4. Repeat step *3* in a similar manner to make up solutions of pH = 3 to pH = 6 and two-thirds fill each appropriate test tube.

5. Two-thirds fill the appropriate test tube with freshly boiled purified water (pH = 7).

6. Two-thirds fill the last test tube with 1·0M sodium hydroxide solution (pH = 14).

7. Prepare solutions of pH = 13 to pH = 8 in a similar manner as given in steps *2* to *4*.

8. To each tube add *two* drops of an indicator and mix by inverting two or three times.

Observations

Using coloured pencils, shade to complete an indicator colour chart:

indicator	colour shading over pH range
	0 1 2 3 4 5 6 7 8 9 10 11 12 13 14
arrange in order of increase of pH change	
universal indicator	

C Construction of a titration curve employing universal indicator

Theory

During a titration the pH of the solution is continuously changing, and this can be conveniently followed by using universal indicator.

Requirements

80 cm³ 0·1M hydrochloric acid
110 cm³ 0·1M sodium hydroxide solution
phenolphthalein indicator solution
B.D.H. universal indicator solution
B.D.H. universal indicator pH colour chart
specified volumetric ware plus a 10 cm³ pipette
100 cm³ graduated flask

Procedure

1. Put the sodium hydroxide solution into the burette and titrate one 20 cm^3 pipetted portion of 0·1M hydrochloric acid, using phenolphthalein as indicator (this provides a check on the equivalence of the molarities).

2. Titrate another 20 cm^3 pipetted portion of 0·1M hydrochloric acid, containing five to seven drops of universal indicator, with 0·1M sodium hydroxide solution, by adding (*a*) three 5 cm^3 portions, (*b*) ten 1 cm^3 portions and (*c*) three 5 cm^3 portions. After the addition of each portion, swirl, and estimate the pH from the colour chart and record its value.

3. Pipette 10 cm^3 of the 0·1M alkali solution into the 100 cm^3 graduated flask. Make up to the mark with purified water and shake well to give 0·01M sodium hydroxide solution.

4. Repeat the titration again as follows:

(*a*) pipette 20 cm^3 of the acid into a conical flask and add five to seven drops of universal indicator,

(*b*) run in the 0·1M alkali solution until 1 cm^3 *less* than that required in step *1* has been added,

(*c*) now add the 0·01M alkali solution, 0·5 cm^3 at a time, and assess the pH after each addition. Continue until 20 cm^3 have been added (equivalent to 2 cm^3 of 0·1M sodium hydroxide solution),

(*d*) record, in cm^3 of 0·1M sodium hydroxide solution, the total volume of alkali added together with its associated pH.

Observations

1. Record all results in a single table as follows:

cm^3 of 0·1M sodium hydroxide added	colour	pH of colour

2. Plot a graph of pH against the volume of 0·1M sodium hydroxide solution added to 20 cm^3 of 0·1M hydrochloric acid.

Question

How does the pH vary when

(*a*) a strong acid is titrated with a weak base,

(*b*) a weak acid is titrated with a strong base,

(*c*) a weak acid is titrated with a weak base?

Sketch graphs of the variation of pH during these titrations.

D *Comparison of acid strengths using the methyl violet indicator chart*

Requirements

40 cm^3 of each of the following acids: 1·0M hydrochloric acid, acetic acid, monochloro-acetic acid and trichloro-acetic acid; 0·5M sulphuric acid and oxalic acid
methyl violet indicator solution
methyl violet indicator pH colour chart
six boiling tubes
six adhesive labels

Procedure

1. Half fill a boiling tube, appropriately labelled, with the acetic acid, add two drops of methyl violet and invert to mix well. Observe the colour shade, estimate the pH from the colour chart and record its value.

2. Repeat step *1* for the other acids.

3. Arrange and list the six acids in order of increasing strength.

Questions

1. Study and complete the information in the following tables:

name of acid	constitutional formula of acid	approximate pH of 1·0M solution
acetic		
monochloro-acetic		
dichloro-acetic		0·7
trichloro-acetic		

name of acid	constitutional formula of acid	approximate pH of 0·5M solution
oxalic		
malonic		1·3
succinic		1·9

name of acid	constitutional formula of acid	approximate pH of 0·5M solution
succinic		1·9
malic		1·5
tartaric		1·3

What explanations may be given to the trends in the pH values for each of the three series of acids?

2. What is the significance of the chosen molarities for the various acids?

3. Why was the pH of dichloro-acetic acid not investigated practically?

4. By what other methods may the strengths of aqueous acid solutions be determined?

E *Investigation of salt hydrolysis using the universal indicator chart*

Requirements

15 cm^3 of each of the following 1·0M salt solutions: ammonium acetate, sodium acetate, zinc nitrate, sodium dihydrogen orthophosphate, disodium hydrogen orthophosphate and trisodium orthophosphate

15 cm^3 of each of the following 0·1M salt solutions: ammonium chloride, magnesium sulphate, sodium bicarbonate and sodium carbonate

B.D.H. universal indicator solution

B.D.H. universal indicator pH colour chart

test tube

Procedure

1. Half fill the test tube with the ammonium acetate solution, add two drops of the universal indicator and invert to mix well. Observe the colour, estimate the pH from the colour chart and record its value.

2. Repeat step *1* for the other salt solutions.

Calculations

Tabulate, as shown below, the 1·0M salt solution in order of increasing pH. Do likewise for the 0·1M salt solutions.

salt solution	molarity	pH	percentage hydrolysis to one significant figure

Two examples for the calculation of percentage hydrolysis are:

(*a*) The primary hydrolysis of copper(II) sulphate is

$$[Cu(H_2O)_4]^{2+}(aq) \rightleftharpoons [Cu(H_2O)_3(OH)]^{+}(aq) + H^{+}(aq)$$

and for a 0·1M solution the pH is 5·5, i.e. $[H^+] = 10^{-5\cdot5}$, which gives the percentage hydrolysis as $\dfrac{10^{-5\cdot5}\times100}{0\cdot1}$, i.e. 0·003.

(*b*) In a 1·0M potassium cyanide solution, the pH is 11·6:

$$[H^+] = 10^{-11\cdot6}.$$

Since

$$[H^+]\,[OH^-] = 10^{-14}$$

$$[OH^-] = 10^{-2\cdot4}$$

and this gives a percentage hydrolysis of $\dfrac{10^{-2\cdot4}\times100}{1}$, i.e. 0·4.

Carry out similar calculations for all the salt solutions. (Where hydrolysis beyond the primary stage takes place, this should be ignored.)

Questions

1. What are the explanatory equations for the hydrolyses in aqueous solution of (*a*) the three orthophosphates and (*b*) the other salts?

2. What happens to sodium chloride when it is made up into a 1·0M aqueous solution, and what is the pH of this solution?

15.2 Choice of indicators for titrimetric analysis

A *Alkali determination by a double indicator method**

Theory

Sodium hydroxide is usually contaminated with sodium carbonate owing to adsorption of atmospheric carbon dioxide.

When sodium hydroxide solution is neutralised with a strong acid,

i.e. $$NaOH(aq)+HCl(aq) \longrightarrow NaCl(aq)+H_2O(l) \qquad (a)$$

there is a sudden and great change in pH, so any indicator can be employed in determining the end point. But the neutralisation of sodium carbonate solution (pH about 11) by a strong acid proceeds in two stages:

$$Na_2CO_3(aq)+HCl(aq) \longrightarrow NaCl(aq)+NaHCO_3(aq) \qquad (b)$$

$$NaHCO_3(aq)+HCl(aq) \longrightarrow NaCl(aq)+CO_2(aq/g)+H_2O(l) \quad (c)$$

with sodium bicarbonate solution (pH about 9) formed as the intermediate. Reaction (*b*) can be followed if an indicator such as phenolphthalein is selected which changes colour within the stated pH limits.

With phenolphthalein as indicator, addition of a strong acid to a solution of sodium hydroxide containing sodium carbonate discharges the colour of the indicator when reactions (*a*) and (*b*) are complete. If methyl orange is then added, further addition of acid changes the colour of the indicator when reaction (*c*) is complete.

As one molecule of sodium bicarbonate is produced from one molecule of sodium carbonate, the volume of acid required for reactions (*b*) and (*c*) will be identical.

Requirements

150 cm^3 0·1M hydrochloric acid
100 cm^3 c. 0·1M solution of sodium hydroxide contaminated with sodium carbonate
phenolphthalein indicator solution
methyl orange indicator solution
specified volumetric ware

Procedure

1. Pipette 20 cm^3 of the alkali solution into a conical flask, add two drops of phenolphthalein, and titrate with the hydrochloric acid until the pink colour is just discharged. Record the volume of acid added.

2. Add four drops of methyl orange and add a further volume of acid to bring about a colour change in the indicator. Again record the volume of acid added.

3. Repeat steps *1* and *2* to obtain concordant results.

Calculations

Determine:

(*a*) the molarity of the sodium hydroxide solution,
(*b*) the molarity of the sodium carbonate impurity,
(*c*) the independent concentrations by weight of sodium hydroxide and sodium carbonate,
(*d*) the percentage purity of the sodium hydroxide.

Question

Giving full theory, how can the percentage composition of a mixture of sodium carbonate and bicarbonate be determined by a double indicator method?

B *Alkali determination by a precipitation method**

Theory

Sodium sesquicarbonate contains both sodium carbonate and bicarbonate, and the total alkali present is estimated, in the normal way, by titration with standard acid using methyl orange as indicator.

The addition of excess sodium hydroxide to the sodium sesquicarbonate solution will convert the sodium bicarbonate to sodium carbonate:

$$NaHCO_3(aq)+NaOH(aq) \longrightarrow Na_2CO_3(aq)+H_2O(l).$$

Then all the sodium carbonate can be precipitated as barium carbonate by the addition of excess barium chloride solution:

$$BaCl_2(aq)+Na_2CO_3(aq) \longrightarrow 2NaCl(aq)+BaCO_3(s)$$

and the excess sodium hydroxide can be found by titrating with standard acid using phenolphthalein as indicator.

Requirements

200 cm^3 sodium sesquicarbonate containing 8·0 g/l
175 cm^3 0·1M hydrochloric acid
100 cm^3 0·1M sodium hydroxide
50 cm^3 c. 0·2M barium chloride solution
phenolphthalein indicator solution
methyl orange indicator solution
specified volumetric ware
test tube

Procedure

1. Titrate 20 cm^3 pipetted portions of the sodium sesquicarbonate solution with 0·1M hydrochloric acid, using methyl orange as indicator to obtain concordant results.

2. Put separate 20 cm^3 pipetted portions of the sodium sesquicarbonate followed by the sodium hydroxide solutions into a conical flask, add half a test tube (c. 10 cm^3) of the barium chloride solution and shake.

3. Add two drops of phenolphthalein and titrate the excess 0·1M sodium hydroxide with the standard hydrochloric acid *slowly*, with constant shaking.

4. Repeat steps *2* and *3* to obtain concordant results.

Calculations

1. For the sodium sesquicarbonate solution, determine:

(*a*) the molarity as a carbonate,

(*b*) the molarity as a bicarbonate.

2. Since the formula of the sesquicarbonate may be written as $x\ Na_2CO_3 \,.\, y\ NaHCO_3 \,.\, z\ H_2O$, calculate:

(*a*) the weights of sodium carbonate, sodium bicarbonate and water of crystallisation in 8·0 g of the solid sesquicarbonate,

(*b*) the simplest ratio of x:y:z.

Question

In step *3* of the procedure, why must phenolphthalein and not methyl orange be used as indicator in the estimation of the excess alkali?

C *Titrimetric investigation of the tribasicity of orthophosphoric acid**

Theory

Orthophosphoric acid, H_3PO_4, is a tribasic acid, and neutralisation may be effected in three stages. In an approximately 0·1M solution the pH values for the monobasic, dibasic and tribasic stages are about 4·6, 9·7 and 12·6 respectively.

For the first stage, bromocresol green (pH range 3·8-5·4) may be used as indicator for titration with standard 0·1M sodium hydroxide solution (use of a comparison solution of the indicator in approximately 0·05M sodium dihydrogen phosphate aids the attainment of the correct end point).

Thymolphthalein (pH range 9·3-10·5) is a suitable indicator for titration with a strong base at the second stage.

The pH change on direct titration at the third stage is negligible and there is no satisfactory indicator. However, if excess silver(I) nitrate solution is added to orthophosphoric acid, all the potential hydrogen ions are progressively liberated,

i.e. $$H_3PO_4(aq) + 3Ag^+(aq) \longrightarrow Ag_3PO_4(s) + 3H^+(aq)$$

during titration with standard sodium hydroxide solution, using bromothymol blue as indicator. Towards the end point the silver(I) orthophosphate precipitate coagulates, and the colour change of the indicator is easily observed in the supernatant liquid.

Requirements

350 cm³ 0·1M sodium hydroxide solution
200 cm³ c. 0·1M orthophosphoric acid
200 cm³ c. 0·1M silver(I) nitrate solution
50 cm³ c. 0·05M sodium dihydrogen phosphate solution
bromocresol green indicator solution
thymolphthalein indicator solution
bromothymol blue indicator solution
specified volumetric ware plus a 10 cm³ pipette, a 250 cm³ conical flask and a white tile
test tube

Procedure

1. Titration as a monobasic acid

(*a*) Pour two test tubes (c. 40 cm³) of sodium dihydrogen phosphate solution into a conical flask and add four drops of bromocresol green. Swirl the solution and set aside the conical flask on a white tile.

(*b*) Pipette 20 cm³ of the orthophosphoric acid into another conical flask standing on a white tile, add four drops of bromocresol green, and run in the standard sodium hydroxide solution until the colour of the comparison solution is matched.

(*c*) Repeat step (*b*) to obtain concordant burette readings.

2. Titration as a dibasic acid

Titrate 10 cm³ pipetted portions of the orthophosphoric acid containing four drops of thymolphthalein, with the standard sodium hydroxide solution to obtain concordant results.

3. Titration as a tribasic acid

(*a*) Pipette 10 cm^3 of the orthophosphoric acid into a conical flask and add two test tubes (c. 40 cm^3) of the silver(I) nitrate solution.

(*b*) Add two drops of bromothymol blue, and with constant shaking run in the standard sodium hydroxide solution. If the colour of the indicator fades towards the end point, add another two drops.

(*c*) Repeat step (*b*) to obtain concordant results, but on approach of the end point allow the precipitate to settle so that the colour change of the indicator may be clearly seen.

Observations and Calculations

1. By reference to (*a*) the pH values for 0·1M solutions and (*b*) the volumes of sodium hydroxide solution employed for the three stages of the neutralisation of orthophosphoric acid, sketch a titration curve of pH against volume of 0·1M sodium hydroxide solution added for the progressive neutralisation of 20 cm^3 of the orthophosphoric acid. Comment on the relationship between the volumes of alkali required for the three stages of neutralisation.

2. Determine the molarity of the orthophosphoric acid.

Questions

1. What are the ionic equations for the three stages of the dissociation of orthophosphoric acid?

2. Why does it become necessary to add more indicator in the presence of the precipitate of silver(I) phosphate?

3. Given that the pH values at the points of equivalence are 8·3 and 3·8 respectively, what is the sketch titration curve for the neutralisation of 20 cm^3 of 1·0M sodium carbonate solution by means of 0·5M sulphuric acid?

16

SOLUBILITY PRODUCT

Introduction

For a saturated aqueous solution of a sparingly soluble strong electrolyte in the presence of the same undissolved solute, the following species exist in equilibrium:

$$\underset{\text{solid}}{MA} \rightleftharpoons \underset{\text{saturated solution}}{MA \rightleftharpoons M^+ + A^-}$$

where MA, whether solid or in saturated solution, represents a single entity of the solute which may be covalent molecules or associated ion pairs, i.e. M^+A^-.

Applying the Law of Mass Action to the over-all equilibrium,

$$K = \frac{[M^+]\,[A^-]}{[MA(s)]}.$$

For this heterogeneous system, with excess solid solute present, and at a given temperature, $[MA(s)]$ is a constant. Hence

$$S = [M^+]\,[A^-]$$

where S is the solubility product of the sparingly soluble electrolyte at the given temperature.

Questions

1. For a sparingly soluble strong electrolyte, M_2B, why is the solubility product expression $[M^+]^2\,[B^{2-}]$?

2. What is the true state of the solute ions in the aqueous solution? What bearing has this on the equilibrium equation and the deduced solubility product expression?

16.1 Determination of the solubility product of copper(II) iodate*

Theory

For a saturated aqueous solution of copper(II) iodate, the solubility product expression for the sparingly soluble strong electrolyte is $[Cu^{2+}]\,[IO_3^-]^2$.

The concentration of a saturated solution may be determined by the addition of excess potassium iodide in the presence of dilute acid. Copper(II) and iodate ions both react quantitatively with iodide ions:

$$2Cu^{2+}(aq)+4I^-(aq) \longrightarrow 2CuI(s)+I_2(aq)$$

$$4IO_3^-(aq)+20I^-(aq)+24H^+(aq) \longrightarrow 12I_2(aq)+12H_2O(l)$$

i.e. $$2Cu^{2+}(aq)+4IO_3^-(aq)+24I^-(aq)+24H^+(aq) \longrightarrow 2CuI(s)+13I_2(aq)+12H_2O(l)$$

to produce iodine, which may then be estimated by titration with a standard solution of sodium thiosulphate.

As a preliminary exercise it is necessary to standardise the sodium thiosulphate solution, and this may be done by using a standard solution of potassium dichromate. A measured volume of a standard solution of potassium dichromate is treated in acid conditions with excess potassium iodide:

$$Cr_2O_7^{2-}(aq)+6I^-(aq)+14H^+(aq) \longrightarrow 2Cr^{3+}(aq)+3I_2(aq)+7H_2O(l).$$

The liberated iodine is titrated with the sodium thiosulphate solution which is to be standardised:

$$I_2(aq)+2S_2O_3^{2-}(aq) \longrightarrow 2I^-(aq)+S_4O_6^-(aq).$$

Requirements

100 cm^3 copper(II) iodate saturated solution
100 cm^3 0·01M potassium dichromate solution
250 cm^3 c. 0·05M sodium thiosulphate solution
50 cm^3 c. 10 per cent potassium iodide solution
50 cm^3 glacial acetic acid
25 cm^3 bench dilute sulphuric acid
starch indicator
specified volumetric ware plus a 250 cm^3 conical flask and a filter funnel
test tube

Procedure

1. Standardise the sodium thiosulphate solution as follows:

(*a*) pipette 20 cm^3 of the standard solution of potassium dichromate into a conical flask, and add a quarter of a test tube (c. 5 cm^3) of both the bench dilute sulphuric acid and the potassium iodide solution,

(*b*) from the burette, run in the sodium thiosulphate solution to titrate the liberated iodine; when the iodine solution reaches a pale straw colour starch may be used as indicator,

(*c*) repeat steps *1* (*a*) and (*b*) to obtain concordant results.

2. Estimate the saturated solution of copper(II) iodate as follows:

(*a*) using a fluted filter paper, filter about 100 cm^3 of the saturated solution into a conical flask,

(*b*) pipette 20 cm^3 of the filtered copper(II) iodate solution into the second conical flask, and add half a test tube (c. 10 cm^3) of glacial acid, followed by a quarter of a test tube (c. 5 cm^3) of the potassium iodide solution,

(*c*) titrate the liberated iodine as in step *1* (*b*),

(*d*) repeat steps *2* (*b*) and (*c*) to obtain concordant results.

Calculations

Determine:

(*a*) the molarity of the sodium thiosulphate solution,

(*b*) for the saturated solution of copper(II) iodate:

(*i*) the molarity,

(*ii*) the ionic concentrations, in moles per litre, of copper(II) ions and iodate ions,

(*iii*) the solubility product of copper(II) iodate at the laboratory temperature.

Question

Why is acetic acid, and not a mineral acid, used in the reactions for the liberation and subsequent titration of iodine from copper(II) iodate and potassium iodide solutions?

16.2 Determination of the solubility product of lead(II) iodide

Theory

For the addition of an aqueous solution of lead(II) ions to an aqueous solution of iodide ions at the point of precipitation of lead(II) iodide, the ionic product

$$[Pb^{2+}]\,[I^-]^2$$

is equal to the solubility product of this sparingly soluble strong electrolyte at the temperature of precipitation.

Requirements

25 cm³ 0·1M potassium iodide solution
250 cm³ 0·01M lead(II) acetate solution
200 cm³ bench dilute sodium hydroxide solution
200 cm³ bench dilute hydrochloric acid
30 cm³ acetone
two 50 cm³ burettes
250 cm³ conical flask
1 cm³ syringe and needle to give drops of c. 0·01 cm³
test tube
black card (c. 15 cm × 15 cm)
hot air blower

Procedure

1. (*a*) Into the conical flask, burette 25·00 cm³ of the lead(II) acetate solution.

(*b*) From the syringe, add 0·25 cm³ of the potassium iodide solution, shake well and then make further additions of one drop at a time; constantly swirl and view, preferably in strong sunlight, against the black card. Record the volume of potassium iodide solution required for the first signs of permanent cloudiness (i.e. the point of precipitation).

(*c*) Discard the contents of the conical flask, and wash with half a test tube (c. 10 cm³) of bench sodium hydroxide solution, followed by an equal volume of bench dilute hydrochloric acid and plenty of purified water. Rinse the conical flask with a quarter of a test tube (c. 5 cm³) of acetone and dry at the hot air blower.

(*d*) Repeat steps (*a*) to (*c*) to obtain concordant readings.

2. Repeat step *1* using the buretted portions of lead(II) acetate and water as indicated in the table below instead of 25·00 cm³ of the lead(II) acetate solution alone.

Observation and Calculations

1. Complete a table identical to that shown on page 228, expressing concentrations in moles per litre.

cm^3 of 0·01M lead(II) acetate solution	25·00	20·00	15·00	10·00	5·00
cm^3 of water	0·00	5·00	10·00	15·00	20·00
cm^3 of 0·1M potassium iodide solution					
total cm^3 at precipitation					
$[Pb^{2+}]$					
$[I^-]^2$					
$\frac{1}{[I^-]^2}$					
$[Pb^{2+}]\,[I^-]^2$ i.e. S					

2. The solubility product expression can be rearranged to give

$$[Pb^{2+}] = \frac{S}{[I^-]^2}$$

and thus a plot of $[Pb^{2+}]$ against $\frac{1}{[I^-]^2}$ should lead to a straight line of slope S. Draw this summarising graph and comment on its form.

Question

Plot the independently determined values of the solubility product against the concentration of the lead(II) ions of the solutions specified in step *1* (*a*) above. What light does this plot throw on the value of the solubility product of lead(II) iodide?

16.3 Determination of the solubility product of calcium hydroxide*

Theory

Calcium hydroxide is a sparingly soluble strong electrolyte, and the solubility product expression, $[Ca^{2+}]\,[OH^-]^2$, is constant no matter whether a saturated solution in water or aqueous caustic alkali (i.e. in the presence of a common ion) is examined. The alkali content of the solutions is estimated by normal acidimetric procedure.

Requirements

400 cm³ 0·1M hydrochloric acid
100 cm³ saturated solution of lime water
100 cm³ 0·025M sodium hydroxide solution saturated with lime
100 cm³ 0·033M sodium hydroxide solution saturated with lime
100 cm³ 0·050M sodium hydroxide solution saturated with lime
100 cm³ 0·100M sodium hydroxide solution saturated with lime
phenolphthalein indicator solution
specified volumetric ware plus a 250 cm³ conical flask and a filter funnel

Procedure

1. Filter about 100 cm³ of the saturated solution of calcium hydroxide through a fluted filter paper into a conical flask, and titrate at least three 20 cm³ pipetted portions of the filtrate with 0·1M hydrochloric acid, using phenolphthalein as indicator.

2. Repeat the same procedure for the other solutions.

Observations and Calculations

1. Complete a table with the following headings:

solution of lime in	cm³ of 0·1M HCl for 20 cm³ of lime solution	ionic concentrations in mole/l				$[Ca^{2+}][OH^-]^2$ i.e. S
		total OH^- i.e. $[OH^-]$	OH^- from NaOH	OH^- from lime	$[Ca^{2+}]$ i.e. half OH^- from lime	

2. (*a*) The solubility product expression can be rearranged to give

$$[OH^-]^2 = \frac{S}{[Ca^{2+}]}$$

and thus a plot of $[OH^-]^2$ against $\frac{1}{[Ca^{2+}]}$ should lead to a straight line of slope S.

(*b*) Tabulate the data for the various solutions:

$[OH^-]^2$					
$1/[Ca^{2+}]$					

(*c*) Draw the suggested graph and determine the value of S from its slope.

Question

What effect has the sodium hydroxide solution on the solubility of calcium hydroxide?

16.4 Titrimetric application of the difference in solubility products of silver(I) chloride and silver(I) chromate

Theory

When silver(I) nitrate is added to a neutral chloride solution containing potassium chromate, silver(I) chloride, being less soluble, is precipitated in preference to silver(I) chromate. (The solubility products for silver(I) chloride and silver(I) chromate are approximately 10^{-10} and 10^{-12} respectively.) Complete precipitation of the silver(I) chloride may be achieved just prior to the critical precipitation of red silver(I) chromate (i.e. the end point of the titration) provided that the potassium chromate is present in the correct concentration.

A *Estimation of a solution of silver(I) nitrate with a standard solution of sodium chloride*

Requirements

0·4 g A.R. sodium chloride
50 cm^3 c. 0·05M silver(I) nitrate solution
5 cm^3 c. 5 per cent potassium chromate solution
specified volumetric ware plus a 10 cm^3 pipette and a 100 cm^3 graduated flask
ignition tube

Procedure

1. Weigh out accurately about 0·29 g of the sodium chloride, dissolve it in purified water and dilute to 100 cm^3 in the graduated flask. Shake well to give a standard solution of about 0·05M.

2. Pipette 10 cm^3 of the sodium chloride solution into a conical flask, add a fifth of an ignition tube (c. 0·75 cm^3) of the potassium chromate solution, and with continuous shaking run in the silver (I) nitrate solution from the burette until a permanent tinge of red appears.

3. Repeat the titration to obtain concordant results.

Calculations

Determine:

(*a*) the molarity of the sodium chloride solution,
(*b*) the molarity of the silver(I) nitrate solution,
(*c*) the concentration of the silver(I) nitrate solution in grams per litre.

Questions

1. What is the correct concentration for the chromate ions in moles per litre at the end point, if all the chloride ions are to be precipitated prior to chromate ions? Use the solubility product values given in the *theory* section.

2. Why cannot potassium chromate be used as indicator in (*a*) acid solution and (*b*) alkaline solution? Investigate qualitatively and explain with the aid of ionic equations.

B *Estimation of a solution of silver(I) nitrate with a standard solution of hydrochloric acid*

Requirements

2·0 g A.R. calcium carbonate
50 cm^3 0·05M hydrochloric acid
50 cm^3 c. 0·05M silver(I) nitrate solution
5 cm^3 c. 5 per cent potassium chromate solution
specified volumetric ware plus a 10 cm^3 pipette
two ignition tubes

Procedure

1. (*a*) Pipette 10 cm^3 of the 0·05M hydrochloric acid into a conical flask. Add with shaking a portion of calcium carbonate which fills an ignition tube to a depth of 1 cm (c. 0·25 g), followed by a fifth of an ignition tube (c. 0·75 cm^3) of the potassium chromate solution.

(*b*) Agitate and titrate with the silver(I) nitrate solution.

2. Repeat step *1* to obtain concordant results.

Calculations

Determine (*a*) the molarity and (*b*) the concentration, in grams per litre, of the silver(I) nitrate solution.

Questions

1. (*a*) What is the purpose of the addition of calcium carbonate, and what is the minimum mass of it required for the success of each titration?

(*b*) Why cannot barium carbonate be used instead of calcium carbonate?

(*c*) What carbonate could be used as an alternative to calcium carbonate?

2. How can the percentage composition of a solution of sodium chloride in hydrochloric acid be estimated?

C *Determination of the number of molecules of water of crystallisation in barium chloride crystals*

Requirements

0·75 g A.R. barium chloride crystals
2·0 g A.R. anhydrous sodium sulphate
50 cm^3 0·05M silver(I) nitrate solution
5·0 cm^3 c. 5 per cent potassium chromate solution
specified volumetric ware plus a 10 cm^3 pipette and a 100 cm^3 graduated flask
two ignition tubes

Procedure

1. Weigh out accurately about 0·6 g of the barium chloride crystals into a 100 cm^3 beaker and add purified water to make a solution. Transfer this solution and washings to the 100 cm^3 graduated flask, make up to the mark with purified water and shake well.

2. (*a*) Pipette 10 cm^3 of this solution into a conical flask. Add a portion of the anhydrous sodium sulphate which fills an ignition tube to a depth of 1 cm (c. 0·4 g), and shake. Pour in a fifth of an ignition tube (c. 0·75 cm^3) of the potassium chromate solution.

(*b*) Titrate with the standard silver(I) nitrate solution.

3. Repeat step *2* to obtain concordant volumes of the silver(I) nitrate solution.

Calculations

Determine:

(*a*) from the titration results, (*i*) the molarity of the solution and (*ii*) the concentration of the solution in grams per litre, (*b*) the percentage of water of crystallisation in the crystals,

(*c*) the number of molecules of water of crystallisation per molecule of barium chloride.

Questions

1. What is the purpose of the addition of sodium sulphate, and what is the stoichiometric mass required for each titration?

2. Why would magnesium sulphate be a suitable substitute for sodium sulphate in the preliminary treatment? What other sulphates could be used in the same way?

APPENDIX A

Page 2

1.1 Make up the alcohol/water mixture in bulk for class use.

Plastic (or glass) cordial bottles are satisfactory for the collection of displaced water.

Page 7

1.3 Employ A.R. potassium hydrogen oxalate as the acid oxalate.

Page 9

1.4 0·03M iron(II) oxalate solution is prepared by dissolving 5·40 g in 3M sulphuric acid and diluting to one litre with dilute sulphuric acid.

Page 10

1.5 The iron(II) oxalate solution is made by dissolving 6·0 g in 3M sulphuric acid and diluting to one litre with dilute sulphuric acid.

Page 11

1.6 The formic acid solution is prepared by diluting 2 cm^3 (dispensed by a 2 cm^3 pipette and pipette filler) of 98/100 per cent formic acid to one litre with purified water.

Page 12

1.7B c. 0·1M alcoholic iodine solution is prepared by dissolving the iodine in 75 per cent alcohol.

Page 16

1.7D Employ A.R. potassium hydrogen iodate, $KIO_3.HIO_3$, as the acid iodate.

Page 17

1.7E The aqueous solution of iodine and iodide ions is prepared by dissolving 6·0 g iodine and 12·0 g potassium iodide in purified water and diluting to one litre.

Page 18

1.8 The experiment is best performed in pairs to cover the alternative mixtures in the third column of the table.

Page 20

1.9 The formaldehyde solution is prepared by diluting about 40 cm^3 of commercial formaldehyde solution (37/41 per cent w/v formaldehyde) diluted to a total volume of about one litre and neutralising to phenolphthalein.

Page 23

1.10 c. 0·05M iodine solution is made by dissolving about 13 g iodine and 30 g potassium iodide in purified water and diluting to one litre.

The bromine water is made by either dissolving about 8·0 g (2·2 cm^3) bromine in one litre of purified water (with the aid of a mechanical stirrer in a fume cupboard), or diluting about 270 cm^3 of a saturated aqueous solution of bromine (which contains c. 30 g/l) to one litre.

Page 24

1.11 Aniline hydrochloride is a suitable alternative to phenol.

Page 26

1.12 Precipitation of other basic salts can be investigated in a similar way, e.g. for zinc, nickel and aluminium.

Page 29

2.1 The ammonium oxalate solution is prepared, at least twenty-four hours prior to use, by adding 50 g to one litre of purified water (solubility is about 4 g per 100 g water at 20°C). Decant the saturated solution just prior to use.

Page 32

2.3 Prepare the sulphur dioxide solution by passing a *slow* stream of sulphur from a canister up through a burette (with a diameter as large as possible) of acetone until the bubbles do not decrease in size on ascending.

Page 37

2.6 The saturated solution of oxygen in water is prepared by shaking a suitable volume of freshly boiled purified water and a volume of oxygen in an enclosed vessel.

Page 38

2.7 The aniline hydrochloride solution is prepared by dissolving about 150 g in about 300 cm^3 of purified water (filter by suction pump if necessary) and diluting to one litre.

Page 40

2.8 Other mixtures which can be investigated are 4-toluidine and 1-naphthol, 4-toluidine and 2-nitrophenol, naphthalene and diphenyl, naphthalene and 4-nitrotoluene, naphthalene and naphthylamine, phenol and urea.

Page 45

3.2 The plastic bottle is used in place of a vacuum flask, and such a replacement is suitable in other thermochemical experiments.

Page 47

3.3 The plastic container No. 8 lid (internal diameter 6 cm) is obtainable from Margros Ltd., Monument House, Monument Way West, Woking, Surrey.

Page 58

4.1 Suitable gases are oxygen and carbon dioxide from cylinders, sulphur dioxide from a canister, picnic stove gas and coal gas.

Page 60

4.2A Suitable liquids are chloroform, carbon tetrachloride and ethyl bromide.

Page 62

4.2B Suitable liquids are diethyl ether, acetone and ethyl alcohol.

Page 64

4.3 The syringe is obtainable from Hospital and Laboratory Supplies Ltd., 12 Charterhouse Square, London E.C.1.

Page 66

5.1 Nitrobenzene may be employed instead of acetamide by dispensing portions up to a maximum of 2 cm^3 from a syringe.

The plastic water container No. 16 is obtainable from Margros Ltd., Monument House, Monument Way West, Woking, Surrey.

Page 94

7.4 The salt solutions in dilute nitric acid are c. 10 per cent.

Page 96

7.5 Supply c. 10 per cent salt solutions, and prepare the two unknown solutions by mixing any two and any three of the four solutions in equal volumes.

Page 98

7.6 Supply c. 10 per cent aqueous solutions of the metal sulphates.

Page 107

.13 Acetic acid may be employed in place of oxalic acid with estimation using standard sodium hydroxide solution.

The adsorption of potassium permanganate on barium sulphate may be investigated in a similar way with estimation using standard oxalic acid solution.

Page 114

.3 Prepare the 'flag' of platinum foil by folding double a 2·5 cm × 2·5 cm piece at the end of a 7·5 cm platinum wire, heating the fold very strongly with a Bunsen flame and repeatedly pinching the hot fold with pliers. After cooling, secure the other end of the wire in a 30 cm glass tube.

Page 118

.8 The saturated solutions of oxalic acid and manganese(II) sulphate are prepared by dissolving 9·5 g and 105·0 g respectively in 100 cm^3 purified water.

Page 120

.10 Due to fluctuations in the laboratory temperature, the 2·5 cm^3 reading (see table, page 121) may take less than three minutes. Should this be so, use 0·5, 1·0 and 2·5 cm^3 0·0001M copper(II) sulphate solution instead of 2·5, 4·5 and 6·0 cm^3 0·001M solution.

Page 129

.2B The outcome of the experiment is not affected if the sodium thiosulphate solution is not accurately 2M.

Page 130

.2C The acidified potassium iodide solution is prepared by dissolving 5·0 g potassium iodide in 100 cm^3 bench dilute hydrochloric acid and diluting to one litre with purified water.

Page 134

.3A The bromine water is prepared either from a saturated aqueous solution diluted about ten times or by dissolving (preferably with a mechanical stirrer and in a fume cupboard) 1·4 cm^3 bromine in 125·0 cm^3 water.

The sodium formate solution is prepared by dissolving 8·0 g in purified water and diluting to one litre.

The 0·001M sodium thiosulphate solution is made by diluting the 0·01M solution.

Page 137

9.3C The borate buffer solution is prepared by dissolving 3·1 g boric acid i. 25 cm³ 1·0M sodium hydroxide solution and diluting to 250 cm³ with purified water.

Page 143

9.3E The methyl formate must be purified *just* prior to use by shaking the required volume with solid sodium bicarbonate, quickly filtering through a fluted filter paper, drying over fused calcium chloride for twenty minutes, decanting the ester and distilling over a water bath (b.p. of ester = 32°C).

Page 144

9.3F The methyl acetate (b.p. = 57°C) is purified just prior to use as in Appendix A, 9.3E.

Page 163

10.5B The copper(I) iodide suspension is prepared as follows: mix 15 cm³ 10 per cent potassium iodide solution with 5 cm³ 5 per cent copper(II) sulphate solution; remove the iodine liberated by running in 0·05M sodium thiosulphate solution from a burette using starch as indicator; wash the precipitate by decantation and suspend it in 100 cm³ purified water.

Page 173

11.6 The best preparation of murexide indicator is a finely ground mixture of murexide with sodium chloride in the mass ratio of 1:100.

Page 180

12.1 The polystyrene spheres for making space-filling models based on atomic covalent radii are obtainable from Elford Plastics Ltd., Brookfield Works, Wood Street, Elland, Yorkshire, and the commercial covalent models from Catalin Products Ltd., Waltham Abbey, Essex.

Page 184

12.2 The alcoholic silver(I) nitrate solution must be prepared with absolute ethanol.

The electrical circuit for testing the conductivity of solutions consists of electrodes of nickel wire mounted to fit a test tube and connected to a 4-volt battery and a 0-6 v. voltmeter.

Page 185

12.3 Tablets of potassium permanganate are obtainable from Risdale and Co. Ltd., Newby, Middlesbrough, Yorkshire.

Employ the same alcoholic silver(I) nitrate solution and electrical circuit as given in Appendix A, 12.2.

Page 189

12.5 For the tablets of potassium permanganate see Appendix A, 12.3

Page 191

12.7 For the tablets of potassium permanganate see Appendix A, 12.3

Page 198

13.2 The distribution of phenoxyacetic acid between chloroform and water may be investigated by a similar method. The phenoxyacetic acid solution is prepared by making a 0·2M solution in chloroform. This may then be titrated, using 0·1M sodium hydroxide solution.

Page 209

14.6 The acetic acid/sodium acetate buffer solution of pH 4 is made by adding 82 cm^3 0·2M acetic acid to 18 cm^3 0·2M sodium acetate, using A.R. chemicals.

Page 213

15.1B Dispense the indicators from polythene bottles with dispenser tops which are obtainable from X-lon Products Ltd., 48 Gillingham St., London S.W.1.

The test tube blocks are best made from blocks of sycamore (40 cm $\times$ 5 cm $\times$ 5 cm) by drilling fifteen holes each with a $\frac{9}{16}$ in diameter and $1\frac{1}{2}$ in deep, and fitted with 100 mm $\times$ 12 mm soda glass test tubes.

Page 218

15.2A Prepare the c. 0·1M solution of sodium hydroxide contaminated with sodium carbonate by dissolving 3·6 g sodium hydroxide and 0·53 g sodium carbonate in purified water and diluting to one litre.

Page 220

15.2B The sodium sesquicarbonate solution is made by dissolving in *cold* water, preferably with the aid of a mechanical stirrer.

Page 221

15.2C c. 0·1M orthophosphoric acid is prepared by diluting 6·5 cm^3 syrupy orthophosphoric acid with purified water to one litre.

Page 225

16.1 The saturated solution of copper(II) iodate is made by adding about 5·0 g to one litre of purified water, heating to about 60°C for a few minutes and allowing the solution to cool to room temperature. This solution must be prepared at least twenty-four hours prior to use and shaken shortly before use.

Page 228

16.3 Prepare the 0·05M, 0·033M and 0·025M sodium hydroxide solutions from 0·1M sodium hydroxide solution by dilution; add calcium hydroxide to each to form a layer about 1 cm thick on the bottom.

These solutions must be shaken regularly over twenty-four hours and immediately prior to use.

With a liberal supply of conical flasks and funnels much time can be saved, as filtrations can be run concurrently.